Coal Spontaneous Combustion Theory with Chain Self-promoted Oxidizing Induced by Active Group

Jun Deng · Yaqing Li · Saeid Zeinali Heris · Chi-Min Shu

Coal Spontaneous Combustion Theory with Chain Self-promoted Oxidizing Induced by Active Group

Springer

Jun Deng
Xi'an University of Science and Technology
Xi'an, Shaanxi, China

Saeid Zeinali Heris
Xi'an University of Science and Technology
Xi'an, Shaanxi, China

University of Tabriz
Tabriz, Iran

Yaqing Li
Xi'an University of Science and Technology
Xi'an, Shaanxi, China

Chi-Min Shu
Department of Safety, Health, and Environmental Engineering
National Yunlin University of Science and Technology
Douliou, Yunlin, Taiwan

ISBN 978-981-96-5361-4 ISBN 978-981-96-5362-1 (eBook)
https://doi.org/10.1007/978-981-96-5362-1

This work was supported by National Natural Science Foundation of China Youth Project (51604218), National Natural Science Foundation of China (51974235) and Shaanxi Provincial Nature Foundation Basic Research Project (2018JM5121).

This Springer imprint is published by the registered company Springer Nature Singapore Pte Ltd.
The registered company address is: 152 Beach Road, #21-01/04 Gateway East, Singapore 189721, Singapore

Preface

Coal remains a vital fossil energy resource. However, the complexity of geological conditions frequently results in disasters during coal mining. Spontaneous coal combustion is one of the most conspicuous hazards, particularly in China. Reports indicate that 70% of mine disasters in China are caused by coal spontaneous combustion. Consequently, spontaneous combustion and its associated secondary disasters severely threaten coal mine safety and production.

Understanding the mechanism of coal spontaneous combustion is fundamental to comprehending its occurrence and progression. However, coal's intricate and highly complex molecular structure has hindered fully understanding this phenomenon. Investigating the oxidation characteristics of coal's active groups and unveiling the essence of spontaneous combustion are of great scientific significance. Such research advances theoretical understanding and facilitates the development of efficient techniques for preventing and controlling coal fires.

Scholars have proposed various theories to explain coal spontaneous combustion, including the bacterial action theory, the pyrite theory, the phenol-based action theory, and the coal-oxygen composite theory. The coal-oxygen composite theory is widely recognised as the most representative explanation, aligning closely with observed phenomena. However, this theory has yet to resolve critical questions, such as which active groups in coal interact with oxygen, the conditions required for such interactions, and the chemical reaction mechanisms involved. These unresolved issues hinder a deeper understanding of the combustion mechanism and limit advancements in fire prevention technologies, creating substantial challenges for on-site fire control in coal mines.

To address these research gaps and limitations, this monograph delves into the microscopic and macroscopic characteristics of coal, drawing on principles from particle physics, free radicals, coal chemistry, structural chemistry, and frontier orbital theory. It introduces solvent extraction assisted by ultrasonication into the study of coal spontaneous combustion, overcoming existing technical bottlenecks. By integrating fundamental principles of particle physics with coal chemistry theories, this research mutually validates findings and enhances systematic investigations into the micro-mechanisms of coal spontaneous combustion. As a result, a novel theory titled *Coal Spontaneous Combustion Theory with Chain Self-promoted Oxidizing Induced by Active Group* is proposed.

This monograph comprises six chapters, focusing on four primary aspects: "*Characteristics of Coal Molecular Structure and Construction of Molecular Models of Active Groups*", "*Ultrasonic Extraction of Active Groups on Coal Molecular Surfaces and Their Oxidation Reaction Characteristics*", "*Molecular Dynamics of Evolution of Active Groups during Coal Oxidation Reactions*" and "*Mechanism of Active Group Oxidation Reactions and Generation Mechanism of Gaseous Products.*" It aims to provide theoretical support for accurately assessing the degree of coal spontaneous combustion hazards and developing advanced inhibitors.

The content elucidates the correlation mechanism between coal active groups and indicator gases, explains the generation mechanisms of indicator gases during coal spontaneous combustion, and lays the foundation for establishing an early warning indicator system. The proposed hydroxyl chain self-promoted oxidation mechanism offers a scientific basis for designing targeted and directional inhibitors.

We anticipate to express our sincere gratitude to all contributing authors. Their invaluable efforts made this monograph possible. We also extend our appreciation to the reviewers for their time and dedication in ensuring the quality of the work, enabling it to meet the high standards of Springer publications.

Special acknowledgment goes to Mr. Xinzu Jin, Mr. Zitao Song, Mr. Haodong Yang, Mr. Zheng Bian, and Ms. Shiyao Hao for their assistance in completing this monograph within a short timeframe.

We are also deeply grateful to Ms. Vinothini Elango (Project Coordinator, Total Service Books Production, Springer Nature) and Mr. Zisheng Wang (Associate Editor, Applied Sciences Springer, part of Springer Nature) for their editorial guidance and seamless collaboration throughout the development of this publication.

We hope readers will find the monograph *Coal Spontaneous Combustion Theory with Chain Self-promoted Oxidizing Induced by Active Group* both insightful and valuable and that it serves as a foundation for further research and innovation in this critical field.

December 2024

Jun Deng
(Responsible for writing Chaps. 3 and 4)
Professor and President
Xi'an University of Science and Technology
Xi'an, China

Yaqing Li
(Responsible for writing Chaps. 5 and 6)
Professor
Xi'an University of Science and Technology
Xi'an, China

Saeid Zeinali Heris
(Responsible for writing Chap. 1)
Professor
University of Tabriz
Tabriz, Iran

Chi-Min Shu
(Responsible for writing Chap. 2)
Distinguished Chair Professor
National Yunlin University of Science and Technology
Douliu, Taiwan

Competing Interests The authors have no competing interests to declare that are relevant to the content of this manuscript.

Contents

List of Figures

List of Tables

Chapter 1
Introduction

Abstract Starting from the position of coal and the distribution of coal resources in the global and Chinese energy structure, this chapter analyzed the major threaten of coal fires for the safety production of coal mines resulted by coal spontaneous combustion, and emphasized the importance and urgency of studying mechanism of coal spontaneous combustion. Advances on mechanism of coal spontaneous combustion including the bacterial action theory, the pyrite theory, the phenol-based action theory and the coal-oxygen composite theory have been analyzed systematically. More importantly, advantages of the coal-oxygen composite theory as well as its unsolved problems such as oxidation pathway of active groups, reaction conditions, microstructure evolution law, etc. were analyzed. Aiming at research bottleneck due to extremely complex of coal molecular structure, this chapter proposed a new idea to investigate mechanism of coal spontaneous combustion by employing ultrasonic extraction technology combined with molecular dynamics simulation method.

1.1 Overview

1.1.1 Distribution of Coal Resources

Coal is one of the world's most important and indispensable non-renewable energy sources, as well as part of China's energy structure. It will continue to maintain its dominant position in global energy production and use for a long time in the future. The concept of green coal resources emerged in the context of the "energy revolution" facing the development of China's coal industry, where coal, as a crucial basic energy source and raw material, strongly supports the rapid development of China's national economy. General Secretary Xi Jinping proposed in the report of the 20th National Congress to "actively and steadily push forward carbon peaking and carbon neutrality" and to "strengthen the clean and efficient use of coal," emphasizing the need to ensure the security of energy resources. Carbon emission reduction in coal and its related industries plays a decisive role in realizing the national goal of "carbon neutrality." This involves achieving coal-based carbon emission reduction

J. Deng et al., *Coal Spontaneous Combustion Theory with Chain Self-promoted Oxidizing Induced by Active Group*, https://doi.org/10.1007/978-981-96-5362-1_1

and promoting the efficient, clean, and low-carbon utilization of coal in the context of carbon peaking and carbon neutrality. It also presents new requirements for the exploration, development, and rational utilization of coal, as well as for improving resource utilization efficiency and alleviating China's shortage of strategic mineral resources. In 2023, the energy sector will promote energy security and a green low-carbon transformation while fully leveraging coal's role as the main energy source. According to statistics, in 2023, raw coal production reached 4.71 billion tons, an increase of 3.4% over the previous year, with coal accounting for 55.3% of China's total energy consumption. Although China's "dual-carbon" policy aims to promote the growth of coal demand at a lower level, it is predicted that by 2050, China's coal consumption will still account for ca. 50%. China's coal is mainly distributed across 17 provinces and regions in the southwest and northeast of the country, with geologically proven reserves of about 28 billion tons. However, due to shallow deposits and the thickness of coal seams in certain coal basins, it is mostly suitable for open-pit mining, making it an important energy supply base.

Southwest China is home to some of the most recent coal resources in the country, with the majority concentrated in Yunnan province. This province has over 30 billion tons of identified resource reserves, with lignite reserves amounting to about 15.5 billion tons, which accounts for more than 50% of the province's total coal resources. In contrast, the total lignite resources in Sichuan province are less than 600 million tons, representing only 5.75% of the province's coal resources, primarily located in the former Xikang province. However, no production mines have been established there yet. Coal in Tibet is only sporadically distributed, and no Neogene coal resources have been found in Guizhou province or Chongqing Municipality.

Coal resources in the south-central region are mainly concentrated in Guangxi, while Hainan and Guangdong provinces have fewer resources. The proven geological reserves of coal in Guangxi are approximately 656 million tons, with the primary coal types being lignite and long-flame coal. Many long-flame coals with corrosive mud are also found in the Nimziping Coalfield, which has an average oil content of 17.3%. However, because it is a non-terrestrial sedimentary coalfield, the sulfur content is high, averaging more than 3% (St,d). The Baise Coalfield, which features the No. 41 coal, also has a high sulfur content, with an average exceeding 3%. While the Baise Coalfield is dominated by No. 41 long-flame coal, the Nanning Coalfield primarily consists of lignite from the Neoproterozoic era, exhibiting a lower average degree of coalification than the Baise coal. In Hainan province, the proven geological reserves of coal exceed 100 million tons, but the coal quality at the Danxian coal mine is poor. In Qiongshan County, there are two mines, the Changchang coal mine, where the thickness of the coal does not exceed 1.50 m. In Guangdong province, the Maoming and Gaozhou Coal fields are present. The proven geological reserves of Maoming coal are just over 16 million tons, while the Gaozhou Shiku coal field has been nearly completely mined out.

Coal in East China is primarily distributed in the Shandong Huangxian coal field, which is currently operated by Longkou coal group. The predicted geological reserves in the land portion are about 1.3 billion tons, with more than 1.25 billion tons of geological reserves expected to be available in the sea under the continental shelf along Huanghai bay. The land mine is accompanied by four layers of oil shale resources, both of which can be mined simultaneously. The Longkou coalfield's Paleoproterozoic coal is predominantly No. 41 long-flame coal. For example, the Liangjia and Beisu mines both produce No. 41 long-flame coal, while the Wulili coal mine contains No. 52 lignite due to its shallow burial. Additionally, the Wutu mine, which has less than 6 million tons of proven geological reserves, produces No. 41 long-flame coal that is characterized by high sulfur content and ultra-high ash content. This coal was formed during the early period when the sea and land intersected.

China's coal resources in northern China are primarily distributed in Inner Mongolia and Hebei province. In Inner Mongolia, the Wuchuan circulation No. well field has proven geological reserves of 117.04 million tons, while the Jining Maliantan mine has proven geological reserves of 238.24 million tons. Currently, both of these sites are only engaged in small-scale kiln mining. In Hebei province, the coal resources are limited to the Zhangbei coalfield, which has proven geological reserves of 48.591 million tons, and the Baoding Laiyuan coalfield, with reserves of 25.733 million tons. The Zhangbei coalfield features 8–9 layers of recoverable coal seams, primarily mining the third coal group, with a recoverable thickness of 6.34 m, indicating a promising future for development and utilization. The Baoding Laiyuan coalfield includes the Doojunwan coal mine, which has a production capacity of 100,000 tons per year and a recoverable thickness of coal seams of about 7.0 m. The reserves of Cenozoic coal in Shanxi province account for an exceptionally small proportion of the total, with only predicted reserves currently being exploited in small kilns. The estimated reserves of Cenozoic coal in Shanxi province are only 475 million tons, primarily distributed in the Pinglu-Recheng and Yuanqu coal production areas, but there is currently no geological exploration data available.

1.1.2 Overview of the Mechanism of Coal Spontaneous Combustion

The mechanism of CSC is fundamental to the study of disaster prevention and control related to this phenomenon. Although the occurrence of CSC was identified as early as the seventeenth century, a complete understanding of its mechanisms remains elusive. For many years, scholars both domestically and internationally have conducted extensive research on the mechanisms of CSC, proposing various theories, including bacterial action (Titilawo et al. 2020), pyrite theory (Xi et al. 2019; Deng et al. 2015), phenolic action (Ma et al. 2018; Fillo 1979), and coal-oxygen composite action (Zhang et al. 2024a, b). Among these, the coal-oxygen composite action theory is broadly accepted by scholars because it posits that CSC is primarily caused by the oxidation of coal and the resultant heat generation. This perspective aligns more closely with the actual phenomena observed in CSC. However, this theory has not adequately addressed several critical questions, such as which functional groups in coal interact with oxygen, the nature of these interactions, the conditions required for these reactions, and the mechanisms of the subsequent chemical reactions. As a result, the inability to explore the mechanisms of CSC in depth has hindered the understanding of its macro–micro correlations. This gap in knowledge has created a bottleneck in the research and development of new technologies for preventing and controlling CSC, complicating efforts to implement effective fire prevention and control measures in coal mines.

The prevailing theory of coal structure is widely accepted as a two-phase model. In this model, the primary component of coal consists of a three-dimensional network of macromolecules, within which small molecules are integrated through hydrogen bonding, electron donor–acceptor interactions, and van der Waals forces. Research has demonstrated that the reactive groups present in coal are the primary factors contributing to the occurrence of spontaneous combustion.

(1) The physical characteristics and chemical structure of coal molecules during spontaneous combustion are often studied independently. Any changes in the physical characteristics observed during CSC are essentially macroscopic manifestations of changes occurring in the chemical structure of coal molecules at the microscopic level (Bodzek and Marzec 1981). Theoretically, there is an inevitable causal relationship between the two; however, existing studies have artificially separated them.
(2) The micro-mechanism of the spontaneous combustion process in coal is primarily based on theoretical analysis and numerical simulations, but it lacks experimental verification. This gap is largely due to the extreme complexity of coal's composition and structure, as well as the mutual conversion of functional groups during the oxidation process (Zheng et al. 2021). Current testing and research methods are unable to distinguish between primary and secondary groups, making it difficult to directly assess the microstructure of the molecules and functional groups involved in the reactions, as well as their oxidation properties in coal.
(3) There is a lack of theoretical guidance and scientific basis for optimizing active group extractants for CSC. Extraction is a physical process that involves dissolving soluble substances within the coal molecular structure in a solvent by disrupting weak interaction forces, such as hydrogen bonding and van der Waals forces, using organic solvents. Theoretically, this process should align with the principles of elementary particle physics. However, existing research in the field of coal chemistry is primarily based on experimental results and lacks a solid theoretical foundation. This not only increases the workload and complexity of extractant screening but also results in a lack of scientific basis for evaluating the effectiveness of the extraction process.
(4) The scientific basis for unitizing the factors contributing to CSC requires further improvement. Given the extreme complexity of coal's structure and the interconversion of various functional groups during the oxidation process, existing testing and research methods are unable to distinguish between primary and secondary groups (Wu et al. 2023a, b). Some scholars have proposed unitizing

the spontaneous combustion factors of coal. However, current unitary research methods often use a single benzene ring to represent the main skeleton of coal when constructing simplified models of coal molecules, rather than relying on the number of atoms in specific aromatic layers as measured by experiments. This approach notably undermines the scientific validity and reliability of the research findings.

1.1.3 Problems with Coal Safety

Coal serves as China's "ballast" for energy security and stable supply. General Secretary Xi Jinping has emphasized that "China is rich in coal but poor in oil and gas," indicating that it is challenging to fundamentally change the coal-dominated energy structure in the short term. Safety is the top priority in the coal mining process, as it is crucial for maintaining national energy security and supply stability. In 2022, the national coal mine mortality rate decreased to 0.054 per million tons, yet there remains a noteworthy gap compared to developed countries, such as the United States and Australia. Approximately 92.6% of China's coal mines are shaft mines, and as the depth and intensity of coal mining increase, the geological conditions become increasingly complex. This complexity poses serious threats, including CSC and the compounded risk of gas disasters. If measures for preventing and controlling CSC and gas disasters are inadequate, there is a heightened risk of severe accidents triggered by gas explosions from spontaneous combustion seams. Furthermore, there is a risk of dust explosions, disruptions in the mine ventilation system, and a series of other serious secondary disasters.

Owing to the complex conditions associated with coal mining, disaster accidents occur frequently. Among these, mine fire accidents rank as one of the five major disasters in China's coal mines, with fires caused by spontaneous coal combustion accounting for 70% of all mine fire incidents (Yu et al. 2022). In addition, spontaneous coal combustion can trigger secondary disasters, such as gas explosion (Liu et al. 2020a), leading to serious casualties and significant economic losses. A notable example is the major accident at the Jilin Babao coal mine on March 29, 2013, which was caused by CSC and a gas explosion in the mining area. This incident resulted in numerous fatalities and substantial economic damage, underscoring the extreme hazards associated with CSC. Statistics indicate that more than 100 workings in state-owned key coal mines in China are closed each year due to spontaneous combustion fires, severely impacting the development of coal resources. Therefore, the prevention and control of spontaneous coal combustion is an urgent issue that must be addressed (Fig. 1.1).

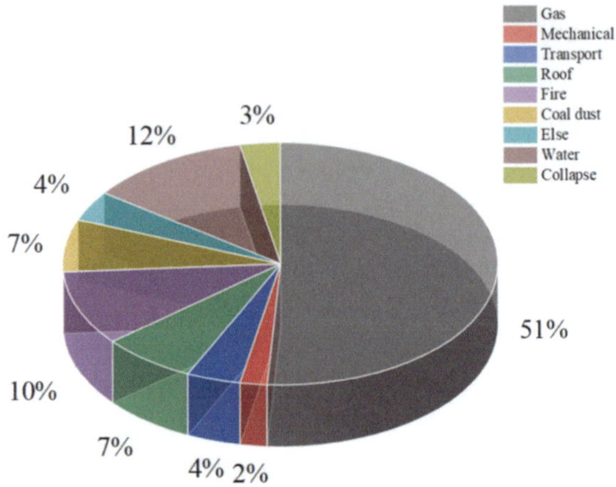

Fig. 1.1 Number of major and particularly serious coal mine accidents in China from 2013 to 2023 (by accident type)

1.2 Coal Spontaneous Combustion Micro Reaction Mechanism Research

To clarify the reaction mechanism of CSC, some scholars have conducted in-depth research on the role of free radicals in this process, the changes in active functional groups during coal oxidation, quantum chemical simulations, and other related features.

1.2.1 Reaction Mechanism of Free Radicals in the Process of Spontaneous Combustion of Coal

The free radical theory of CSC suggests that the coal-oxygen complex results from a reaction between oxygen molecules and active groups (free radicals) on the surface of coal. Hu et al. (2022a, b) investigated the free radical properties of the coal oxidation process under a gas atmosphere using electron paramagnetic resonance spectroscopy and characterized the free radical activity with the Lund factor g. Yang (2021) examined the effect of thermal damage on the changing characteristics of free radicals during CSC, with results indicating that thermal damage can widen the reaction channel and increase reactivity. Zhou et al. (2020, 2021) analyzed the effect of different oxygen concentrations on the reactivity of trace groups in the low-temperature oxidation process of coal. Their results showed that reactive groups were more likely to undergo oxidation reactions at high oxygen concentrations ($O_2 > 9\%$). Cai et al. (2021) noted that an increase in the absolute content of radical concentration could not fully characterize the oxidative properties of coal and proposed that the spontaneous combustion propensity of coal should be characterized in terms of the relative rate of generation of radical concentration. Li et al. (2023a, b) employed electron paramagnetic resonance spectroscopy to test free radical parameters in coal samples and compared the evolution of free radical parameters and various types of free radical concentrations during low-temperature oxidation and low-temperature thermal decomposition using split-peak fittings of free radical profiles. Li et al. (2023a, b) used quantum chemistry to investigate the generation mechanism of peroxyl radicals in coal. Li et al. (2022) explored the influence of different oxygen contents in mine airspace on the changes in free radicals during the low-temperature oxidation process of coal. Their results indicated that coal bodies with different metamorphic degrees can undergo free radical reactions in an oxygen-free atmosphere; as oxygen content increases, the line width of lignite rises, while that of bituminous coal and anthracite decreases; and the relative growth rates of the g-factor, the concentration of free radicals, and the concentration of free radicals in the three types of coal all increase.

1.2.2 The Study of the Change Rule of Active Functional Groups in the Process of Coal Oxidation

The process of CSC is a self-accelerating warming process in which different functional groups are gradually activated in a sequential manner and react with oxygen. This process is influenced by the oxidation activity of the main functional groups in the coal, the conditions under which the reaction occurs, and the ambient temperature.

Lv et al. (2022) found that an increase in the proportion of nitrogen in the atmosphere during coal CSC can lead to changes in the evolutionary characteristics of functional groups in coal. Zhao et al. (2020) studied the microscopic characteristics of methane emissions during CSC under varying oxygen concentrations, particle sizes, metamorphic degrees, and warming rates using Fourier Transform Infrared Spectrometer. They identified the key reactive groups responsible for the generation of methane gas products. Deng et al. (2023) pointed out that CH_4 gas can slow down the process of low-temperature oxidation in coal by inhibiting the content of key reactive groups. The inhibitory effect of CH_4 on coal low-temperature oxidation becomes significant as CH_4 concentration increases within the range of 0–4%. The degree of inhibition on reactive groups, in order of magnitude from largest to smallest, affects oxygen-containing functional groups, aromatic hydrocarbons, aliphatic hydrocarbons, and hydroxyls. Yuan (2022) found that the content of hydroxyl functional groups decreased and then increased with the deepening of the deterioration degree of coal samples; aliphatic hydrocarbon functional groups increased and then decreased; oxygenated functional groups gradually decreased, while aromatic hydrocarbon functional groups gradually increased. The content of reactive groups was higher in coals with a low degree of deterioration. Hou et al. (2022) used DSC and FTIR to investigate changes in exothermic parameters and functional group content during CSC. They found that the ignition point of coal samples containing more –C=C– and substituted benzene groups was higher, indicating a lower risk of spontaneous combustion. Conversely, coal samples with high –COO– and $–CH_2–$ groups exhibited a higher risk of spontaneous combustion. Yin et al. (2024) were the first to use the Kubelka–Munk equation and the Starink method to calculate the functional group content of coal samples, as well as the oxidation kinetics of functional groups. The results showed that the main functional groups in coal, including –OH, –C=O–, $–CH_2–/–CH_3$, –C=C–, and –CH, exhibited different oxidation characteristics at each stage. Miao et al. (2023) tested the evolution of reactive functional groups during coal oxidation using gas chromatography and in-situ infrared spectroscopy, finding that the content of oxygen-containing functional groups gradually increased by 2–3 times with rising temperature. Duan et al. (2023) employed in situ electron spin resonance and in-situ infrared spectroscopy to appraise the changes in the main functional groups during coal oxidation, demonstrating that phenolic hydroxyls, as electron-donating groups, were converted to –CO functional groups through dehydrogenation and oxidation reactions, promoting the formation and accumulation of persistent free radicals in coal.

1.2.3 Research on the Spontaneous Combustion Mechanism of Coal Based on Quantum Chemistry

Quantum chemical calculation is a vital tool for analyzing the structure of coal molecules. Beginning with Heitler and London's study of the structure of hydrogen molecules in 1927, quantum chemical calculations have become intertwined with various subdisciplines of chemistry, as well as physics, biology, and computational mathematics. They have been widely applied in fields such as biology, materials science, energy, and the chemical industry. Currently, many scholars, both domestically and internationally, are using quantum chemical methods to study the mechanisms of CSC. Wang et al. (2024a, b, c) employed quantum chemical methods to assess the transition metal ions in coal samples, demonstrating the existence of active sites, and conducted relevant experiments and simulations to confirm that these transition metal ions promote CSC. Jin (2021) applied quantum chemistry software to analyze the active structure of the superoxide dismutase center in coal molecules, constructed a model of coal peroxyl radicals, and performed calculations to investigate the nature of the inhibition of CSC by superoxide dismutase. Wang and Liang (2020) established a simple molecular structure representing the functional groups of coal, employed Gaussian software for simulations, and combined these with thermogravimetric experiments to study the changes in functional groups during the natural processes of coal. They also analyzed the activity of these functional groups. Huang (2018a, b) investigated the amount of CO produced by CSC at different temperatures and elucidated the mechanism of CO formation using experimental and simulation methods based on quantum chemistry and free radical theory. They proposed that the oxidation of low molecular compounds in coal represents the optimal time to inhibit spontaneous combustion. Gao et al. (2023) employed quantum chemical calculations in conjunction with experiments, such as programmed temperature-programmed analysis, to assess the exothermic intensity of water valley bituminous coal at different stages, demonstrating the feasibility of quantum chemical calculations for evaluating the exothermic intensity of spontaneous coal combustion. Hou et al. (2022) used glutathione as an inhibitor to mitigate the low-temperature oxidation of coal. Through thermogravimetric analyses of coal samples, in-situ FTIR experiments, and quantum chemical calculations using density functional theory, they concluded that the mechanism of glutathione inhibition involves inactivating reactive radicals in coal and converting them into more stable compounds. Zhu et al. (2023) revealed the inhibition mechanism of Tripeptidyl Aminopeptidase I (TPPI) using DFT. Coal samples treated with TPPI and composite corrosion inhibitors exhibited a decrease in the content of aliphatic functional groups, hydroxyl groups, and carboxyl groups, while the relative contents of stable ether bonds and aromatic functional groups increased. This study provided theoretical support for understanding the microscopic mechanism by which TPPI inhibits CSC. Qi et al. (2016) appraise the active centers and reaction pathways of hydroxyl groups during CSC using quantum chemical calculations. They selected Ar–CH_2–$CH(CH_3)$–OH as the structural unit for their study. The results indicated that there are two active centers in this structural unit: The hydrogen atom in the hydroxyl group and the hydrogen atom in the R_3–CH structure. The hydrogen atom in the hydroxyl group is directly extracted by oxygen, while the hydrogen atom in R_3–CH is first extracted by oxygen to form a hydrogen peroxide radical, after which the hydrogen atom in the hydroxyl group is extracted. The study also assessed the microstructural and kinetic parameters of the reactions, revealing that the oxygen extraction reaction occurs at medium temperatures and has a pronounced thermal effect, while the R_3–CH reaction occurs at lower temperatures but provides an active structure for subsequent reactions at higher temperatures. These findings contribute to a deeper understanding of the self-heating mechanism of coal and offer guidance for the utilization and processing of coal to improve efficiency and reduce energy consumption. Wang et al. (2014, 2016a, b) analyzed the active sites in the coal structure, including oxygen-hydrogen bonding, carbon-hydrogen bonding, and the frontier orbitals of radical single electrons. They constructed the active structural units involved in the spontaneous combustion of coal using frontier orbital theory and quantum chemical calculations. During CSC, the radical structural units react with oxygen to release gaseous products, revealing a low activation energy chain cycle in the kinetic process of coal oxidation. The theory proposes a reaction mechanism for the reactive structures and establishes a reaction sequence that includes 13 radical reactions. It also determines the degree of difficulty and the order of occurrence of these reactions, thus elucidating the reaction mechanism of CSC. Zhang et al. (2017) investigated 8 possible radical types and their structures in coal, assessing their stability when in contact with O_2 at room temperature. They found that the radicals generated by breaking the dangling bonds on the aryl ring and alkyl chain are highly reactive to oxidation and form stable oxygen-containing radicals. Through single heavy state-occupied molecular orbital analysis, it was discovered that the radicals produced by O_2 addition on the alkyl group are carbon radicals and that breaking the aryloxy C–O bond produces carbon-centered radicals. It was also found that mobile hydroxyl groups may trigger chain radical oxidation reactions in low-rank coals. Sun et al. (2004) used molecular mechanics and semi-empirical quantum chemistry to study the molecular structure models of Shenmu coal microcomponents, comparing the energy compositions, bond lengths, and bond cleavage energies of the molecular models of the mirror and inert groups. The results showed that torsional and van der Waals energies are the main forces in the molecules and that substituents have a significant effect on the system energy, with alkyl substituents increasing the system energy and phenyl substituents decreasing it.

Although quantum chemical calculations have been applied in the microanalysis of spontaneous coal combustion, they cannot calculate the dynamic properties of coal during the oxidation process. The transition state intermediates and reaction steps obtained from these studies are difficult to verify experimentally. While FTIR can be used to analyze the types and changes of functional groups in coal, it does not allow for the verification of the reaction steps derived from quantum chemical analysis. Therefore, it is challenging to reveal the spontaneous combustion mechanism of mixtures with unclear structures, such as coal, using this instrumental calculation method alone.

1.3 Study of the Molecular Structure of Coal

The properties of coal are determined by its molecular structure, which serves as the foundation for coal chemistry research (Jiang et al. 2023; Wang 2006). Since the beginning of the twentieth century, researchers in coal chemistry have conducted numerous studies on the structure of coal.

Zhang et al. (2023) combined elemental analysis, infrared spectroscopy, and other relevant analyses of lignite to determine its unitary molecular formula as $C_{45}H_{47}O_{12}N$. This model has been widely used in related research. Chen et al. (2023) utilized infrared spectroscopy to analyze coal samples, demonstrating that the coal molecule consists of aromatic, aliphatic, and cyclic basic structural units. Li et al. (2023a, b) concluded that the proportion of aromatic structures in the coal macromolecular structure gradually increases with maturity. Jia et al. (2022) employed X-ray photoelectron spectroscopy and other tests on the microcrystalline structure of coal, concluding that aromatic hydrocarbons were predominantly benzene ring disubstituted, accounting for 44.93% of the total aromatic hydrocarbons, supplemented by benzene ring trisubstitution or pentasubstitution, with the least amount being benzene ring IV. Cui et al. (2020) used 13C-NMR and FTIR to analyze the coal structure and found that ether bonds and methoxy groups were not present in the coal molecular structure. Wang et al. (2024a, b, c) derived that the degree of aromaticity and the condensation of aromatic rings increased with the degree of coalification. Sharma et al. (2024) identified the specular group as the major organic component of coal using XRD, 13C-NMR, and FTIR. Jia et al. (2024), through characterization methods, such as FTIR, XPS, and 13C-NMR, showed that the proportion of oxygen-containing functional groups in coal molecules gradually decreased with increasing deterioration, while the length of long aliphatic chains decreased and aromaticity increased. Li et al. (2024a, b) constructed five models using molecular simulation techniques to study the adsorption and diffusion of H_2S, deriving the effects of different structural defects on these processes. Jiao et al. (2024) analyzed the alcoholysis of bituminous coal using GC/MS, revealing that the proportion of aromatic hydrocarbons in the soluble phase (SP) was the largest at 27.9%, followed by phenolic compounds, indicating that compounds containing benzene rings were more readily extracted during isopropanol alcoholysis. Zhang et al. (2024a, b) employed NMR carbon spectroscopy, infrared spectroscopy, X-ray photoelectron spectroscopy, and other tests to analyze the sequences of 13 different coal macromolecular models of anthracite (with reflectance values ranging from 0.46 to 3.21%) from various coal orders. Li et al. (2023a, b) characterized the coal structure using X-ray diffraction and FITR, combined with split-peak fitting to obtain the relative content of functional groups and structural parameters. Wang et al. (2024a, b, c) determined the elemental composition and structure of Tingnan coal using 13C-NMR, XPS, and FTIR. Wu et al. (2023a, b) tested 19 coal samples using electron paramagnetic resonance (EPR) to clarify their metamorphic characteristics. Ge et al. (2023) constructed a macromolecular structure unit model of coal before and after microwave irradiation. Huang et al. (2023) proposed a molecular structure model for coal rich in northern Shaanxi through experimental characterization tests and theoretical calculations, verifying the model using quantum chemistry, 13C-NMR, and FTIR.

1.4 Research on the Macroscopic Characteristics of Coal Spontaneous Combustion

The performance of the CSC process at the macro level is primarily divided into two aspects: Heat release and the escape of index gases. Consequently, both domestic and international scholars have analyzed the mechanism of CSC from these two perspectives.

1.4.1 Research on Gas Products in the Process of Coal Spontaneous Combustion

Hou et al. (2024) analyzed the gaseous product release behavior of CSC during the low-temperature oxidation stage. They elaborated on the microstructural changes characteristic of the CSC process and predicted the changes in gaseous products during the high-temperature stage through temperature-programmed experiments and in-situ FITR. They concluded that the critical temperature and the dry cracking temperature of CSC were 50 and 90 °C, respectively, and that the released carbon, oxygen gases, and hydrocarbon gases exhibited a parabolic increase with rising temperature. Tian et al. (2024) explored the mechanisms by which gas influences the spontaneous combustion process of coal. Their study included the driving dilution effect of CH_4 on O_2, the displacement effect caused by the competitive adsorption and desorption of CH_4 and O_2, as well as the molecular chemical adsorption of O_2 and the coal-oxygen composite effect in high-temperature environments. Zhao et al. (2020) conducted temperature-programmed experiments on six coal samples with varying degrees of oxidation, plotting the absolute production curves of oxidized gas (O_2) and each index gas (CO, CO_2, C_2H_6, C_2H_4, C_3H_8, C_4H_{10}) as temperature increased. Li et al. (2020a, b) designed a method to obtain the gas production characteristics of the pure oxidation pathway, employing in-situ FTIR and modeling compounds to investigate the relationship between changes in reactive groups and the CO_2/CO ratio during the pure oxidation process. The experimental results indicated that the CO_2 produced by coal at low temperatures was primarily primary CO_2. Their new method could eliminate the effects of other factors, revealing that the CO_2/CO ratio produced by the pure oxidation pathway exhibited distinct stages, which could be divided into four phases as temperature increased. The participation of different reactive groups in the coal determined the initial temperature range of each stage. Xu (2023) utilized temperature-programmed experiments and electron spin resonance spectroscopy to study the effect of gas on the oxidation characteristics of CSC. The results indicated that the generation of signature gas products at the same temperature decreased with an increase in gaoeous volume fraction, and the macroscopic characteristics of spontaneous combustion oxidation exhibited a notable hysteresis phenomenon. Liang et al. (2022) conducted temperature-programmed experiments on eight different coal samples to accurately determine the critical temperature of CSC. They analyzed the relationship between gaseous products and the temperature during the oxidation process of CSC to identify the indicator gases suitable for each stage of combustion.

1.4.2 Study on the Thermal Effect of Coal-Oxygen Reaction in the Process of Coal Spontaneous Combustion

Li et al. (2020a, b) utilized a C_{80} micro-calorimeter to test the exothermic properties of coal during low-temperature oxidation under varying oxygen concentration conditions. They found that there are two critical oxygen concentrations in the experimental coal samples, and the initial exothermic temperature of coal low-temperature oxidation tends to rise as oxygen concentration decreases. Zhang et al. (2022) noted that CH_4 gas in an oxidizing atmosphere curtails the amount of heat absorbed and released during the low-temperature oxidation process. They observed that higher CH_4 concentrations correspond to a lower amount of heat released. Zhang et al. (2021) also employed a C80 micro-calorimeter to study the exothermic properties of coal in a low CH_4 concentration atmosphere. They pointed out that as CH_4 concentration increases, the decrease in heat absorption of coal is <10 J, while the decrease in exothermic heat is close to 300 J. Lv et al. (2021) conducted TG/DSC experiments to investigate the heat loss characteristics of pre-oxidized coal and the relationship between thermal effects, oxidizing temperature, and oxygen concentration. Their findings indicated that pre-oxidized coal exhibits poor ignition performance and burnout capacity. Although its burning rate and short-term burning intensity are higher than those of raw coal, the maximum exothermic power and exothermic capacity of pre-oxidized coal are lower than those of raw coal.

Scholars have combined macroscopic experiments with simulation software to investigate various factors affecting the coal oxidation process. Shi et al. (2021a, b) established a one-step global numerical model of coal oxidation influenced by multi-component substances based on key parameters from temperature-programmed experiments. They studied the changes in the oxygen consumption rate at each stage of the coal oxidation process and identified the key parameters influencing this rate. In addition, they developed a one-step global dynamic model of the chemical reaction between coal and oxygen, considering the effects of multiple substances. Their findings indicated that controlling storage temperature conditions is beneficial for preventing and managing the early-stage oxidation reactions of coal. Furthermore, they discovered that altering oxygen supply conditions can effectively control the later stages of CSC development (Shi et al. 2021a, b).

1.5 Research on Ultrasonic Extraction of Active Functional Groups of Coal Spontaneous Combustion

The ultrasonic extraction technique accelerates the dissolution of extractables through the cavitation effects of ultrasound in the liquid, along with its additional benefits. Wei et al. (1998) extracted coal using a carbon disulfide/N-methyl-2-pyrrolidone (CS_2-NMP) mixture under ultrasonic irradiation at room temperature. They combined this method with high-pressure chromatography to identify the structures of 11 compounds present in the coal. Li et al. (2010) also performed ultrasonic extraction of Lingwu coal using CS_2-NMP mixed solvents, achieving favorable extraction results. Hua et al. (2011) investigated the mechanism of ultrasonic action on small molecules in coal, proposing that the impact flow generated by cavitation can enhance mass transfer between the solvent and coal samples. This process allows for the release of compounds that traditional solvent extraction methods cannot extract within a short time frame. Barma (2018) focused on optimizing the coal beneficiation process through ultrasonic-assisted technology, conducting a mechanistic analysis of the modifications to the coal structure and examining the factors affecting the extraction efficiency of ultrasonic-assisted extraction of coal molecular groups. Lv et al. (2018) suggested that, compared to a single-wave frequency, a dual-frequency ultrasonic coupling can create a more uniformly distributed acoustic field, which alleviates the standing wave effect while enhancing the cavitation effect in the liquid. Guo et al. (2021) analyzed the dominant stages of different microactive structures during the CSC reaction. They summarized the working principles of extraction reagents for targeted extraction of coal molecular microstructures and elaborated on the modification effects of various extraction-assisted techniques on coal structure. Zhang (2014) applied four solvents: acetone, methanol, THF, and DMF to extract Shengli lignite using the ultrasonic-assisted extraction method. They compared the extraction effects of the four solvents, finding that the extraction rates followed the order: DMF > THF > methanol > acetone. Shi et al. (2022) conducted ultrasonic extraction of low-temperature coal tar using an equal volume mixture of petroleum ether and methanol. They experimentally determined the composition of low-temperature coal tar from the Yulin area in northern Shaanxi, China. Huang (2018a, b) utilized seven different extractants and performed extraction experiments on coal samples using both single-solvent and composite-solvent methods with ultrasonic assistance. They compared and analyzed the extraction effects of various extractants on the main active functional groups in the coal samples. In addition, they investigated the oxidative properties of several representative extracted coal samples and original coal samples using a TG/DSC-FTIR coupling system.

Hu et al. (2022a, b) selected eight organic solvents for the extraction of medium-temperature bitumen at room temperature. They obtained the infrared structure of the extract and its apparent activation energy through infrared spectroscopy and thermogravimetric analysis. They established a fitting relationship model between the two, revealing details about the pyrolysis characteristics of coal asphalt and other related information. Liu et al. (2020a, b) employed an experimental orthogonal design method to determine the optimal conditions for ultrasonic extraction of coal using methanol. They comparatively analyzed the effects of temperature, solid–liquid ratio, and extraction time on the results. Yu et al. (2020) used ultrasonic extraction to study the extraction rates of the same coal sample with different extractants and analyzed the chemical structure of the resulting extracts. Arauzo et al. (2020) performed ultrasonic-assisted extraction of dried aqueous char and slurry fractions, selecting three solvents: Water, methanol, and aqueous-methanol mixtures. Their results indicated that ultrasonication-assisted extraction of aqueous char using methanol as the solvent achieved the highest total phenolic content. They also found that milder delignification process conditions resulted in the highest phenolic content being extracted from the aqueous char. Zhou et al. (2020) extracted Victory lignite using a toluene/ethanol mixture under ultrasound-assisted conditions. They further separated the extracts by column chromatography and analyzed 22 esters using gas chromatography/mass spectrometry (GC/MS). The isolation and analysis of these high-value organic compounds provide a crucial theoretical basis for studying lignite as a non-fuel resource. Ju (2021) performed ultrasound-assisted extraction of DX coal using CCl_4 with an ultrasonic cleaner. She explained and analyzed why chlorine-containing compounds were the main products in the solvent extraction process from the perspective of the extraction mechanism. Chen et al. (2018) utilized anhydrous ethanol as a solvent for graded ultrasonic extraction at room temperature, demonstrating that ultrasonic extraction in a dissolved system prominently accelerates the rate of solvent extraction. Zhou (2014) isolated high-purity phosphorus-containing compounds from Shengli lignite extract using ultrasonic extraction for the first time, marking a pronounced discovery in the field of coal chemistry. Yao et al. (2009) employed petroleum ether as the solvent for ultrasonic extraction, showing that all components in kerosene measurable by gas chromatography/mass spectrometry could be effectively separated. Liu et al. (2008) used water for ultrasonic extraction of coal and found that the extraction rate of coal treated with ultrasonic water extraction was greater than that of the raw coal in a mixed solvent. Ma (1990) compared Soxhlet extraction with ultrasonic extraction, demonstrating that ultrasonic extraction is easier and faster, making it suitable for routine analysis of liquefaction products. Liu (2016) revealed the composition and structural characteristics of organic solutes from lignite through a five-stage extraction process using ultrasonic radiation at room temperature and graded variable-temperature hot fusion. Yang (2020) found that the efficiency of extraction increased with the flatness of the solvent structure during ultrasound-assisted extraction. Gao et al. (2016) examined ultrasonic extraction methods and concluded that ultrasonic temperature had the least effect on extraction results compared to other extraction conditions. Wang et al. (2016a, b) conducted ultrasonic extraction experiments on coal samples using a petroleum ether/acetone solvent mixture and analyzed the probable structures of soluble small molecules in coal. Wang (2018) performed ultrasonic radiation extraction of Xinjiang Piliqing sub-bituminous coal using an equal volume mixture of acetone (ACE) and CS_2, followed by heat bath operation. They compared and analyzed the final product with the hot-bath product of the original coal, revealing the compositional and structural characteristics of the soluble organic matter in sub-bituminous coal, thereby laying a solid foundation for improving the efficiency of sub-bituminous coal utilization.

References

Arauzo P, Lucian M, Du L et al (2020) Improving the recovery of phenolic compounds from spent coffee grounds by using hydrothermal delignification coupled with ultrasound assisted extraction. Biomass Bioenergy 139:105616. https://doi.org/10.1016/j.biombioe.2020.105616

Barma S (2018) Ultrasonic-assisted coal beneficiation: a review. Uitras Sonochem 50:15–35. https://doi.org/10.1016/j.ultsonch.2018.08.016

Bodzek D, Marzec A (1981) Molecular components of coal and coal structure. Fuel 60(1):47–51. https://doi.org/10.1016/0016-2361(81)90030-2

Cai J, Yang S, Zheng W et al (2021) Dissect the capacity of low-temperature oxidation of coal with different metamorphic degrees. Fuel 292:120256. https://doi.org/10.1016/j.fuel.2021.120256

Chen GX, Wei CL, Zhang ZY et al (2018) Ultrasonic extraction of Xinjiang Changyan coal with ethanol and analysis of their extract. Mod Salt Chem Ind 45(01):3–10

Chen JL, Zhu WY, Chang MJ et al (2023) Experimental study on flotation of coal slurry from Nanliang coal mine with kerosene/methyl oleate compounding traps. Conserv Utiliz Miner Resour 43(02):20–26

Cui XQ, Wang YQ, Li QH (2020) Construction and analysis of coal molecular model for 1000 MW boiler in thermal power plant. Energy Energy Conser 12:114–182
Deng J, Ma X, Zhang Y et al (2015) Effects of pyrite on the spontaneous combustion of coal. Int J Coal Sci Technol 2:306–311. https://doi.org/10.1007/s40789-015-0085-y
Deng J, Zhou JM, Bai ZJ et al (2023) Effect of gas on microstructure and thermal reactivity of coal during low temperature oxidation. Coal Sci Technol 51(01):304–312
Duan Z, Zhang Y, Deng J et al (2023) Effect of molecular structure change on the properties of persistent free radicals during coal oxidation. Fuel 128:350. https://doi.org/10.1016/j.fuel.2023.128861
Fillo JP (1979) Understanding of phenolic-compound production in coal-gasification processing. Environmental Research and Technology, Inc., Pittsburgh. https://doi.org/10.3390/app13179754
Gao PQ, Qiao ZL, Zhang Y et al (2016) Ultrasonic extraction of phenolic compounds from light oil of medium and low temperature coal tar. Clean Coal Technol 22(2):54–63
Gao F, Bai QH, Jia Z et al (2023) Exothermicity of coal during low temperature oxidation process based on quantum chemical calculation. J China Coal Soc 48(09):3428–3440
Ge T, Wu M, Xu J et al (2023) Effect of 915-MHz low-temperature microwave irradiation on the molecular structure of coking coal. Asia-Pacific J Chem Eng 18(5):e2963. https://doi.org/10.1002/apj.2963
Guo J, Jin Y, Zheng XZ et al (2021) Research and prospect of coal spontaneous combustion inhibition mechanism based on extraction technology. Coal Sci Technol 49(03):92–99
Hou QY, Zhai XW, Song BB et al (2022) Correlation analysis of coal micro-groups and spontaneous combustion risk. Saf Coal Mines 53(04):13–19
Hou YT, Zhang YC, Sun JQ et al (2024) Study on the response characteristics between gas indicators of spontaneous combustion and organic functional groups of coal. China Coal 50(01):67–74
Hu X, Yu Z, Cai J et al (2022a) The influence of methane on the development of free radical during low-temperature oxidation of coal in gob. Fuel 330:125369. https://doi.org/10.1016/j.fuel.2022.125369
Hu ZS, Xu YL, Zhu HY et al (2022b) FTIR analysis of the correlation between the pyrolysis characteristics and molecular structure of ultrasonic extraction derived from mid-temperature pitch. Spectr Spect Anal 42(03):889–895. https://doi.org/10.3964/j.issn.1000-0593(2022)03-0889-07
Hua ZQ, Qin ZH, Chen DR et al (2011) GC/MS analysis of clarain from tongting bituminous coal extracted using soxhlet extraction and ultrasonic extraction. Guangzhou Chem Ind 39(10):1–4. https://doi.org/10.3969/j.issn.1001-9677.2011.10.001
Huang SH (2018a) Formation mechanism of CO of coal spontaneous combustion based on quantum chemistry and radical about on temperature. Electr Power Technol Environ Protect 34(04):18–21
Huang Y (2018b) Ultrasonic extraction and exothermic characteristics of active groups during coal spontaneous combustion process. ShaanXi. https://doi.org/10.27397/d.cnki.gxaku.2018.000045
Huang Z, Zhou W, Wei J (2023) Study on the molecular structure model of tar-rich coal and its pyrolysis process. J Mol Struct 13:1286. https://doi.org/10.1016/j.molstruc.2023.135613
Jia JZ, Xing YH, Li B (2022) Analysis of molecular structure characteristics of anthracite in Shanxi Yangquan coal. Chem Res Appl 34(10):2311–2320. https://doi.org/10.3969/j.issn.1004-1656.2022.10.006
Jia J, Wang D, Li B (2024) Study on molecular structure characteristic and optimization of molecular model construction of coal with different metamorphic grade. J Mol Struct 1295:136655. https://doi.org/10.1016/j.molstruc.2023.136655
Jiang B, Huang J, Yu C et al (2023) Experimental and theoretical study on molecular structure construction of Hongliulin coal. Fuel 349:128708. https://doi.org/10.1016/j.fuel.2023.128708
Jiao KP, Zhao ZT, You CY et al (2024) Composition of the alkanolyses soluble portion and pyrolytic products distribution of the insoluble portion from Wucaiwan sub-bituminous coal. Chem Ind Eng Prog 43(05):2449–2462

Jin BX (2021) Study on reaction mechanism of antioxidant inhibiting coal peroxyradical based on quantum chemical. Tianjin University of Science and Technology, Tianjin. https://doi.org/10.27360/d.cnki.gtlgy.2021.000573

Ju CX (2021) Separation and analysis of organic constituents. Zaozhuang Bituminous Coal, Nanjing. https://doi.org/10.27623/d.cnki.gzkyu.2020.001786

Li Y, Zong ZM, Huang LL et al (2010) Separation and analysis of the extracts in CS2/NMP mixed solvent from Lingwu coal. J Wuhan Univ Sci Technol 33(1):88–94. https://doi.org/10.3969/j.issn.1674-3644.2010.01.019

Li J, Lu W, Liang Y et al (2020a) Variation of CO_2/CO ratio during pure-oxidation of feed coal. Fuel 262:1–13. https://doi.org/10.1016/j.fuel.2019.116588

Li QW, Ren LF, Ren SJ (2020b) Study on low-temperature oxidation characteristics of coal under oxygen deficient atmosphere. Saf Coal Mines 51(11):34–38

Li JH, Yang SQ, Jiang XY (2022) Effect of oxygen content on free radicals during low-temperature oxidation of coal. China Sci 17(09):1043–1048. https://doi.org/10.3969/j.issn.2095-2783.2022.09.015

Li HT, Zou XY, Zhang WG et al (2023a) Molecular evolution of medium rank coal in Southern Shaanxi. J xi'an Univ Sci Technol 43(06):1118–1127

Li W, Yang WB, Zhan XY et al (2023b) Graphitization mechanism of coal organic macromolecular carbon structure. J China Coal Soc 48(02):855–868

Li JY, Li YX, Bai G et al (2024a) Influence of structural defects of tectonically deformed coal on adsorption and diffusion characteristics of hydrogen sulfide. Coal Conver 47(04):1–10

Li JH, Huang JJ, Lu W et al (2024b) Correlation between highly active carbon-containing solid free radicals and spontaneous combustion reactivity of coal. Coal Sci Technol 62:1–20

Liang CC, Li Y, Sang MM (2022) Mathematical statistical analysis of quantitative indicators for coal spontaneous combustion prediction. Coal Chem Ind 45(06):106–109

Liu CM, Zong ZM, Wei XY et al (2008) Effect of water treatment under ultrasonic irradiation on the extraction yields of Dongshen coal in the CS2/NMP mixed solvent. J xi'an Univ Sci Technol 1:32–35

Liu T, Lin B, Fu X et al (2020a) Modeling air leakage around gas extraction boreholes in mining-disturbed coal seams. Process Saf Environ Protect B 141:202–214. https://doi.org/10.1016/j.psep.2020.05.037

Liu H, Yu SY, Li JW et al (2020b) Best condition of ultrasonic extraction design of Xinjiang sulphur ditch coal with orthogonal method. Mod Salt Chem Ind 47(01):34–35

Liu J (2016) Ultrasonic radiation extraction and fractionated thermal dissolution of Zhaotong Lignite, Nanjing

Lv BT, Chen J, Wang Q et al (2018) Soft measurement study of Chinese medicine extraction based on an embedded system. J Beijing Univ Chem Technol 45(03):95–100

Lv H, Deng J, Li D et al (2021) Effect of oxidation temperature and oxygen concentration on macro characteristics of pre-oxidised coal spontaneous combustion process. Energy 227:120431. https://doi.org/10.1016/j.energy.2021.120431

Lv ZG, Xu YL, Liu ZJ et al (2022) Groups evolution of long-flame coal oxidation process under various air ambience. Saf Coal Mines 53(2):46–52

Ma JS (1990) Ultrasonic extraction-a new method for rapid separation of coal liquefaction products. Coal Anal Utiliz 02:5–9

Ma W, Han Y, Xu C et al (2018) Enhanced degradation of phenolic compounds in coal gasification wastewater by a novel integration of micro-electrolysis with biological reactor (MEBR) under the micro-oxygen condition. Bioresour Technol 251:303–310. https://doi.org/10.1016/j.biortech.2017.12.042

Miao G, Li Z, Yang J et al (2023) Microstructure evolution and higher-molecular-weight gas emission during the low temperature oxidation of coal. Energy 12:282. https://doi.org/10.1016/j.energy.2023.128424

Qi X, Xue H, Xin H et al (2016) Reaction pathways of hydroxyl groups during coal spontaneous combustion. Can J Chem 94(5):494–500. https://doi.org/10.1139/cjc-2015-0605

Sharma R, Kumar S, Adak A et al (2024) A comparative study of molecular structure and combustion behavior of coal and its separated vitrains. J Mol Struct 1299:137152. https://doi.org/10.1016/j.molstruc.2023.137152
Shi X, Zhang Y, Chen X et al (2021a) Effects of thermal boundary conditions on spontaneous combustion of coal under temperature-programmed conditions. Fuel 295:120591. https://doi.org/10.1016/j.fuel.2021.120591
Shi X, Zhang Y, Chen X et al (2021b) Numerical study on the oxidation reaction characteristics of coal under temperature-programmed conditions. Fuel Process Technol 213:106671. https://doi.org/10.1016/j.fuproc.2020.106671
Shi H, Wang FQ, He YF et al (2022) Composition analysis of ultrasonic extraction of low temperature coal tar. Sci Technol Chem Ind 30(01):22–24
Sun QL, Li W, Chen HK et al (2004) Molecular modeling of coal macerals by using quantum chemistry. J Fuel Chem Technol 03:282–286. https://doi.org/10.3969/j.issn.0253-2409.2004.03.006
Tian FC, Jia DX, Chen MY et al (2024) Research progress of spontaneous combustion of coal containing gas under the compound disaster environment in the goaf. J China Coal Soc 49(06):2711–2727
Titilawo Y, Masudi WL, Olawale JT et al (2020) Coal-degrading bacteria display characteristics typical of plant growth promoting rhizobacteria. Processes 8(9):1111. https://doi.org/10.3390/pr8091111
Wang BJ (2006) Study on quantum chemistry of coal structure and reactivity. Taiyuan University of Technology, Taiyuan
Wang HW, Liang ZQ (2020) Mechanism Research on Quantum Chemical Reaction of -CH_2-FunctionGroup in Coal Spontaneous Combustion. Energy Technol and Manag 45(06):136–138. https://doi.org/10.3969/j.issn.1672-9943.2020.06.049
Wang DM, Xin HH, Qi XY et al (2014) Mechanism and relationships of elementary reactions in spontaneous combustion of coal: the coal oxidation kinetics theory and application. J China Coal Soc 39(08):1667–1674
Wang D, Xin H, Qi X et al (2016a) Reaction pathway of coal oxidation at low temperatures: a model of cyclic chain reactions and kinetic characteristics. Combust Flame 163:447–460. https://doi.org/10.1016/j.combustflame.2015.10.019
Wang DX, Zhen XQ, Zhao XK et al (2016b) Ether and analysis of their extract ultrasonic extraction of Xinjiang Changyan coal with acetone/petroleum. Guangzhou Chem Ind 44(22):71–86. https://doi.org/10.3969/j.issn.1001-9677.2016.22.026
Wang F, Tan B, Zan X et al (2024a) Molecular simulation study of microstructural evolution during low-temperature oxidation of coal. Energy 290:130054. https://doi.org/10.1016/j.energy.2023.130054
Wang FS, Sun W, Zhang Y et al (2024b) Quantum chemical calculation of coal spontaneous combustion mechanism promoted by transition metal ions. J China Coal Soc 49(05):2347–2359
Wang C, Zeng F, Li C et al (2024c) Insight into the molecular structural evolution of a series of medium-rank coals from China by XRD, Raman and FTIR. J Mol Struct 1303:137616. https://doi.org/10.1016/j.molstruc.2024.137616
Wang YN (2018) Ultrasonic radiation extraction and parallel thermal dissolution of Piliqing subbituminous coal, Nanjing
Wei X, Zong Z, Wu L (1998) LC/MS analysis of CS2/NMP soluble fraction from upper freeport coal. In: Prospects for coal science in the 21st century. Shanxi Science and Technology Press, pp 263–266
Wu J, Wu D, Xia G et al (2023a) Evolutionary analysis of coal molecular structure under different metamorphism: an attempt to use electron paramagnetic resonance. ACS Omega 8(43):40547–40560. https://doi.org/10.1021/acsomega.3c05367
Wu K, Yao Q, Chen Y, Zhao P et al (2023b) Dependence evaluation of factors influencing coal spontaneous ignition. Energy Sci Eng 11(10):3738–3750. https://doi.org/10.1002/ese3.1549

Xi P, Ma R, Liu W (2019) Research on the effect of carbon defects on the hydrophilicity of coal pyrite surface from the insight of quantum chemistry. Molecules 24(12):2285. https://doi.org/10.3390/molecules24122285

Xu YQ (2023) Determination and prediction of the state of coal spontaneous combustion based on multi-objective grey target decision. China University of Mining and Technology. https://doi.org/10.27623/d.cnki.gzkyu.2023.002570

Yang Y (2020) Ultrasonic assisted extraction of different macerals from Yangchangwan Coal by three kinds of organic solvents. Ningxia University, Ningxia. https://doi.org/10.27257/d.cnki.gnxhc.2020.000382

Yang K (2021) Study on the effect of thermal damage on coal structure and spontaneous combustion oxidation characteristics. China University of Mining and Technology, Jiangsu. https://doi.org/10.27623/d.cnki.gzkyu.2021.001581

Yao T, Zong ZM, Yuan NH et al (2009) Isolation and GC/MS analysis of a high-temperature coal tar collected from Xuzhou Coking Plant, Saint-Gobain Pipelines (Xuzhou) Co., Ltd. J Wuhan Univ Sci Technol 32(06):648–653

Yin Y, Zhang Y, Huang Z et al (2024) A kinetic research method for coal oxidation based on the Kubelka-Munk equation and the Starink method. Combust Sci Technol 196(2):245–260. https://doi.org/10.1080/00102202.2022.2072685

Yu SY, Liu H, Li JW et al (2020) Dedign the best condition of ultrasonic extraction of Xinjiang Piliqing coal with orthogonal method. Mod Salt Chem Ind 47(02):42–43

Yu YN, Zhao QW, Cheng M et al (2022) Prevention and control techniques against coal spontaneous combustion under the influence of "breathing effect" in isolated island working face of extra-thick coal seam. J xi'an Univ Sci Technol 42:76–82

Yuan Y C (2022) Research on microstructure change characteristics and macroscopic performance of coal samples with different metamorphic degrees during spontaneous combustion. Inner Mongolia University of Science and Technology, Inner Mongolia. https://doi.org/10.27724/d.cnki.gnmgk.2022.000479

Zhang M (2014) Solvent extraction of Shengli lignite and structure characterization of its small molecular compounds, Nanjing

Zhang Z, Kang Q, Yun T et al (2017) Density functional theory investigation of possible structures of radicals in coal undergoing O_2 chemisorption at ambient temperature. Energy Fuels 31(1):953–958. https://doi.org/10.1021/acs.energyfuels.6b03072

Zhang YT, Zhang YB, Li YQ et al (2021) Thermal effect of coal oxidation and evolution properties of key functional groups of coal oxidation at low gas atmospheres. J China Univ Min Technol 50(4):776–783

Zhang Y, Zhang Y, Li Y et al (2022) Low-temperature oxidation characteristics of coals at the methane-containing atmosphere. J Therm Anal Calorim 147(4):3379–3389. https://doi.org/10.1007/s10973-021-10735-0

Zhang TH, Cheng MY, Tian FC et al (2023) Molecular simulation on competitive adsorption characteristics of $CH_4/O_2/N_2$ gas in lignite. J Min Sci Technol 8(06):817–827

Zhang C, Zhang SH, Tang SH et al (2024a) Construction and comparison of macromolecular models of different coal rank coals. Geoscience 21:1–21

Zhang F, Liu W, Qin Y et al (2024b) Numerical investigation of gas emission and extraction under the action of coal-oxygen recombination in longwall gobs. Gas Sci Eng 124:205258. https://doi.org/10.1016/j.jgsce.2024.205258

Zhao J, Wang T, Deng J et al (2020) Microcharacteristic analysis of CH_4 emissions under different conditions during coal spontaneous combustion with high-temperature oxidation and in situ FTIR. Energy 209:118494. https://doi.org/10.1016/j.energy.2020.118494

Zheng H, Li Y, He W et al (2021) Study on the effect of organic sulfur on coal spontaneous combustion based on model compounds. Fuel 289:119846. https://doi.org/10.1016/j.fuel.2020.119846

Zhou J (2014) Separation and analysis of soluble organic matter from Shengli Lignite and Xinjiang Changyan Coal

Zhou B, Yang S, Wang C et al (2020) The characterization of free radical reaction in coal low-temperature oxidation with different oxygen concentration. Fuel 262:116524. https://doi.org/10.1016/j.fuel.2019.116524

Zhou B, Yang S, Jiang X et al (2021) The reaction of free radicals and functional groups during coal oxidation at low temperature under different oxygen concentrations. Process Saf Environ Protect 150:148–156. https://doi.org/10.1016/j.psep.2021.04.007

Zhu H, Xie L, Huo Y et al (2023) Investigation of the physicochemical inhibitors on coal spontaneous combustion based experimental and quantum chemical methods. J Therm Anal Calorim 149(3):935–951. https://doi.org/10.1007/s10973-023-12778-x

Chapter 2
Characterization of Coal Molecules and Their Surface Structures

Abstract Focusing on basic structural characteristics of coal molecules, basic parameters of bituminous coal and anthracite including vitrinite reflectivity, industrial analysis and elemental analysis were analyzed in detail. The study of this chapter would provide an important basis for the subsequent chapters. At the same time, 13C Nuclear Magnetic Resonance spectroscopy (NMR) and Fourier Transform Infrared Spectroscopy (FT-IR) were employed to analyze the distribution of aromatic structure, fat structure and active functional groups in coal molecules in detail. Relative contents of different aromatic carbons in coal were determined, and ratios of aromatic bridge carbon to weekly carbon were calculated. Furthermore, structural characteristics of coal molecules were revealed. Moreover, aromatic structural units of coal molecules were characterized, which provided basic data support for the subsequent construction of coal spontaneous combustion active group model and the study of its weak interaction with extractants.

The chemical structure of coal molecules is the basis for the study of the oxidative spontaneous combustion process of coal and the spontaneous combustion mechanism of coal (Zhu et al. 2023), and the chemical bonds and functional groups in the organic molecules of coal play a crucial role in the spontaneous combustion process of coal. The rudimentary parameters of the coal samples of bituminous coal (hereafter referred to as bituminous coal for convenience of designation) from the No. 8 coal seam of Datong Coal Yukou Mine and anthracite coal (hereafter referred to as anthracite for convenience of designation) from the No. 29 coal seam of Ningxia Yangchangwan Mine were analyzed through the experiments of specular group reflectance, industrial analysis and elemental analysis on the selected coal samples. Without destroying the coal samples, the distribution characteristics of the reactive groups in the coal molecules were obtained using nuclear magnetic resonance (13C-NMR) to make the constructed aromatic skeleton of coal molecules scientific. Fourier infrared spectroscopy (FT-IR) was used to appraise the distribution characteristics of reactive groups in coal molecules, which provides theoretical basis for the construction of coal molecular model.

J. Deng et al., *Coal Spontaneous Combustion Theory with Chain Self-promoted Oxidizing Induced by Active Group*, https://doi.org/10.1007/978-981-96-5362-1_2

2.1 Characterization of Basic Parameters of Coal Samples

2.1.1 Coal Rock Microcomponent Reflectance

The microscopic components of coal can be divided into three major groups, namely, the vitrinite group, the exinite group and the inertinite group (Bhatta and Dash 2023). The determination of specular group reflectance was based on the national standard "Vitrinite Reflectance Microscopy Method for Coal" (GB/T 6948-2008) and "Determination of Microcomponent Groups and Minerals of Coal" GB/T 8899-2013, and the reflectance of vitrinite and the content of vitrinite were determined by using Kexiang MAC Smart Scope 2000 series coal rock analyzer. The reflectance of vitrinite is a key physical parameter, with the increase of coal body metamorphism degree, the distribution area of vitrinite reflectance will show an obvious trend of widening, and gradually move to the higher reflectance area, this law is helpful to infer the coal body metamorphism degree through the reflectance of vitrinite.

The reflectance of bituminous coal and anthracite is listed in Table 2.1, the reflectance R_{max} of anthracite is 2.83%, and the reflectance R_{max} of bituminous coal is only 0.76%, which indicates that the degree of metamorphism of anthracite is higher, and the degree of metamorphism of bituminous coal is lower.

2.1.2 Industrial and Elemental Analysis of Coal Samples

Proximate analysis and ultimate analysis of coal are the basic contents of coal quality analysis, which are closely related to the cause of coal, the degree of coalification and petrographic composition. The industrial analysis of the coal samples was carried out according to the national standard "Methods of Proximate Analysis of Coal" (GB/T 212-2008), and was measured by the GF-8000B automatic industrial analyzer of Beijing Tiandishouhe. The ultimate analysis of the coal samples were performed according to the national standards "Methods for the Determination of Carbon and Hydrogen in Coal" (GB/T 476-2008), "Methods for the Determination of Nitrogen in Coal" (GB/T 19227-2008) and "Methods for the Determination of Whole Sulfur in Coal" (GB/T 214-2007) respectively, and were completed by Elementar's Vario EL Cube elemental analyzers (CHNS measurement mode) and HTDL-8000 fully automatic sulfur meters. CHNS measurement mode and HTDL-8000 automatic

Table 2.1 The vitrinite group reflectance

Sample	Reflectivity $(R^2_{\max\,\%})$	Average random reflectance $(R^2_{\mathrm{ran}\%})$
Bituminous coal	0.76	0.72
Anthracite coal	2.83	2.51

Table 2.2 Proximate and ultimate analysis of coal samples

Sample	Proximate analysis (%)				Ultimate analysis (%)				
	M_{ad}	A_{ad}	V_{ad}	FC_{ad}	C	H	O	N	S
Bituminous coal	5.37	15.78	15.92	62.93	71.43	3.055	3.443	0.302	0.62
Anthracite coal	1.57	3.46	8.96	86.01	88.56	3.428	2.594	0.268	0.12

sulfur meter of Elementar (Germany). The results of proximate analysis and ultimate analysis of each coal sample are displayed in Table 2.2.

From the analysis of the results obtained in Table 2.2, it can be seen that the moisture content in the two types of coal samples, bituminous coal and anthracite, is in the range of 1.57–5.37%. The ash content of anthracite is lower, 3.46%. The ash content of coal is related to the coal-forming environment and coal-forming process because it originate from the combustion residues of minerals and combustible components in coal (Zhang 2015). The volatile matter Vad content of the two coal samples measured ranged from 8.96 to 15.92%, and the volatile matter content of the coals decreased as the degree of metamorphism increased.

The results of the ultimate analysis of the correlation comparison coal samples showed that among the detected elements, both coal samples are dominated by C element, with an average C content of roughly 78%, and the O element content of the coal samples averages at 3.02%. The S content of anthracite was lower, only 0.12%, and bituminous coal had a higher S content of 0.62%. The above analyses showed that with the increasing degree of coalization, the organic matter in the coal shows obvious characteristics of carbon enrichment, dehydrogenation and deoxygenation of the aromatic nucleation condensation reaction.

2.2 Structural Characteristics of the Main Body of the Coal Molecule

2.2.1 Nuclear Magnetic Resonance Test Instrument and Parameter Setting

A Bruker AVANCE NEO 400M NMR instrument was employed for the NMR carbon spectroscopy test (^{13}C nuclear magnetic resonance spectroscopy, abbreviated as ^{13}C-NMR). The experimental test setup is shown in Fig. 2.1. The resonance frequency of the ^{13}C detection NMR was 15 kHz, the probe was 3.2 mm, the relaxation time was 3 s, the resonance frequency was 15 kHz, the adoption time was 17 ms, the pulse width was 2.4 μs, the cycle time was 5 s, the contact time was 10 ms, and the number of scans was 32.

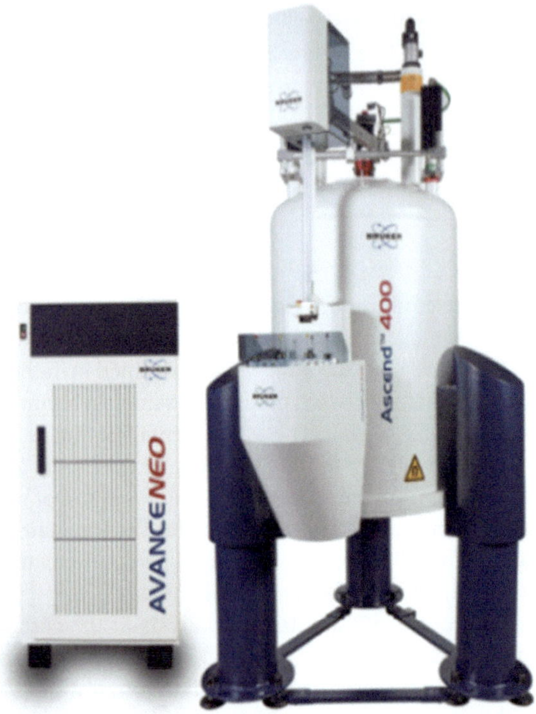

Fig. 2.1 AVANCE NEO 400 M nuclear magnetic resonance machine

2.2.2 *Characterization of the Carbon Structure of Coal Molecules*

Although ^{13}C-NMR spectra can determine the carbon skeleton information of coal, their sensitivity is lower than that of hydrogen spectra due to their low signal-to-noise ratio. Therefore, in order to make the generated ^{13}C-NMR spectra more accurate, cross-polarization (CP) and rotational sideband suppression (TOSS) techniques were used to improve the sensitivity of the spectra when acquiring the ^{13}C-NMR spectra of the coal samples (Chang et al. 2023), and the ^{13}C-NMR spectra of the two types of coal samples are shown in Fig. 2.2.

According to the ^{13}C-NMR spectra of each coal sample in Fig. 2.2, there are three chemical shift peaks in the measured coal molecules, which are categorized as located near 0–60, 60–90, and 90–200 ppm. According to the results of the study obtained by previous researchers (Qin 2019), the individual carbon chemical shifts in the coal molecules are attributed as shown in Table 2.3.

To achieve a better fitting effect, the ^{13}C-NMR spectra of the two coal samples were smoothed and cumulatively peak-fitted using Peakfit software, and the fitted data with a better overlapping effect with the original data were received, and the results of the fitted data are shown in Fig. 2.3.

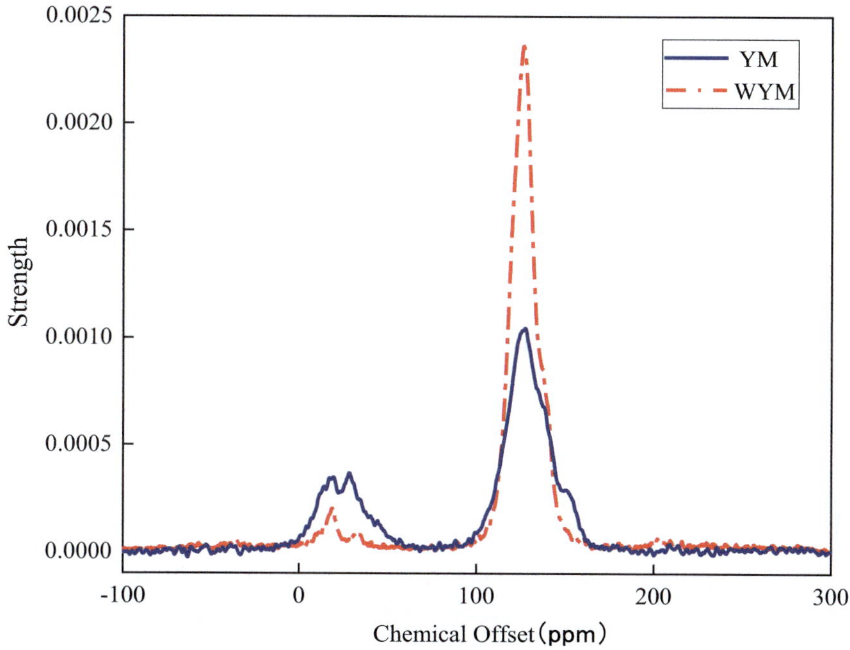

Fig. 2.2 ^{13}C-NMR spectrum of raw coal sample

Table 2.3 ^{13}C-NMR functional group attribution and carbon chemical shift attribution table

Chemical offset (ppm)	Primary attribution	Chemical offset (ppm)	Primary attribution
16	Aliphatic methyl group	100–129	Protonated aromatic carbon
20	Arylmethyl	129–137	Bridged aromatic carbon
23–50	Methylene, hypomethyl	137–148	Side-branched aromatic carbon
50–65	Methoxy and oxygenated methylene carbon	148–165	Oxygen-substituted aromatic carbons
60–70	Oxygenated hypomethyl carbon	165–220	Carboxy and carbonyl carbons
75–90	Oxygenated aliphatic carbon in the ring		

According to the peak attribution of the groups, the area of each attributed peak was integrated, and the percentage of information parameters of 13 chemical structure carbon spectra split-peak fits were calculated for total aromatic carbon, carbonyl and carboxylic carbon, SP^2 hybridized carbon in the aromatic ring, non-protonated

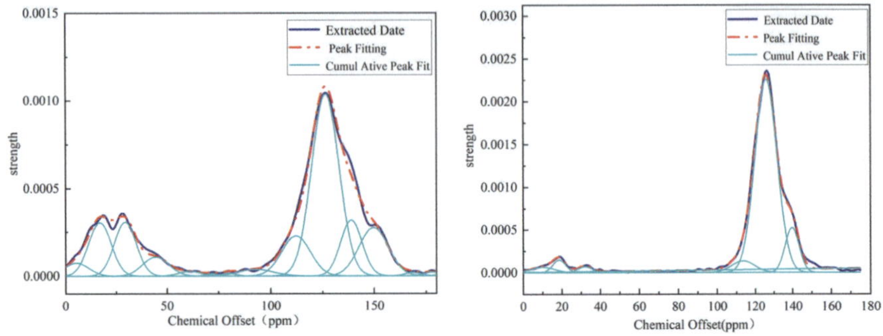

(a) Bituminous coal 13C-NMR peak fitting spectra

(b) Anthracite coal 13C-NMR peak fitting spectra

Fig. 2.3 Fitted ^{13}C-NMR peaks of raw coal samples

carbon, protonated carbon, etc., for both coal samples, and the specific results are showned in Table 2.4.

The results in Table 2.4 showed that the carbon element in coal exists in various forms, most of which are in the form of aromatic rings. Comparing the forms of

Table 2.4 Fitting parameters of the split peaks of ^{13}C-NMR carbon spectra of coal samples

Samples	Bituminous coal/%	Anthracite coal/%
f_a	68.50	81.83
f_a^c	3.36	3.47
$f_{a'}$	65.14	71.16
f_a^N	22.05	24.27
f_a^H	43.05	57.56
f_a^P	1.16	2.12
f_a^S	6.28	1.95
f_a^B	14.61	20.20
f_{al}	26.69	18.17
f_{al}^*	5.10	1.32
f_{al}^H	13.79	10.17
f_{al}^O	7.804	2.24

Note All parameters are in %; f_a—Total aromatic carbon; f_a^c—Carbonyl and carboxylic carbon; $f_{a'}$—SP^2 hybridized carbon in aromatic ring; f_a^N—Non-protonated carbon; f_a^H—Protonated carbon; f_a^P—Replacement of aromatic carbon; f_a^S—Replacement of aromatic carbon; f_a^B—Bridging carbon; f_{al}^*—Total fat carbon; f_{al}—Aliphatic methyl carbon and aromatic methyl carbon; f_{al}^H—CH and CH_2 groups carbon; f_{al}^O—Carbon-oxygenated fat-bonded; $f_a = f_{a'} + f_{ac}$; $f_{a'} = f_a^H + f_a^N$; $f_a^N = f_a^P + f_a^S + f_a^B$; $f_{al} = f_{al}^H + f_{al}^* + f_{al}^0$; $f_a + f_{al} = 100\%$

carbon in coal samples with different degrees of metamorphism, it can be found that the total aromatic carbon (f_a) content of bituminous and anthracite coal was 68.5 and 81.83%, respectively, among them, the total aromatic carbon (f_a) content of anthracite was significantly higher compared to that of bituminous coal, while the corresponding substituted aromatic carbon (f_a^S) was 6.28 and 1.95%, and the lipocarbon rate (f_{al}) was only 26.69 and 18.17%, respectively. As the deterioration degree of the coal samples increased, the total aromatic carbon (f_a) showed an obvious increasing trend, while the substituted aromatic carbon (f_a^S) and lipid carbon rate (f_{al}) gradually decreased. This means that the higher the deterioration degree of the coal samples, the more carbon atoms in the coal molecules existed in the aromatic rings, the number of their side chains decreased pronouncedly, and the aromatization degree of the holostic coal molecules increased gradually. Comparing the connection forms of carbon atoms on the side chains, it can be found that the proportion of –CH and $-CH_2$—group carbon (f_{al}^H) of the two types of coals is 13.79 and 10.17% in turn, while the proportion of aliphatic methyl carbon and aromatic methyl carbon (f_{al}^*) was 5.10 and 1.32%, which means that as the degree of metamorphism increases, the carbon atoms on the side chains of coal molecules were more inclined to be connected with hydrogen atoms, and the main forms of their existence were chain alkanes and cycloalkanes.

Based on ^{13}C-NMR mapping, the relative content of these different aromatic carbons in coal can be determined, the ratio of aromatic bridge carbon to peripheral carbon (referred to as bridge-to-carbonratio) of the samples was calculated according to Equation $X_{BP} = f_a^B / (f_a^H + f_a^P + f_a^S)$, and the specific calculation results are shown in Table 2.5.

According to Table 2.5, the ratio of aromatic bridge carbon to pericarbon was measured to be 0.289 and 0.337 for bituminous coal and anthracite, respectively, which showed that the X_{BP} rises with the increase of the degree of deterioration. It was known that the X_{BP} value of naphthalene was 0.25, and the X_{BP} value of anthracene was 0.40, in which the X_{BP} of bituminous coal and anthracite is between 0.25 and 0.40, and it can be inferred that the molecular skeleton of bituminous coal and anthracite mainly possesses naphthalene as the core structure. In addition, bituminous and anthracite coal samples contain a minute amount of anthracene, phenanthrene and other condensed aromatic ring components, and combining with the relevant literature, it can be seen (Li et al. 2022; Xiang et al. 2016) that when the reflectance of the specular group of the coal was between 0.5 and 3%, the number of

Table 2.5 Ratio of aromatic bridge carbon to pericarbon in coal sample

Simple compound		Aromatic nuclei in coal		
Typology	Ratio of aromatic bridge carbon to pericarbon	Serial number	Ratio of aromatic bridge carbon to pericarbon (X_{BP})	Major type
Bicyclic	0.25	Bituminous coal	0.289	2 and 3 rings mainly
Tricyclic	0.40	Anthracite coal	0.337	2 and 3 rings mainly

aromatic rings in the molecule was 2–4. The mirror group reflectance of bituminous coal and anthracite was 0.76 and 2.83%, respectively. Combining the above two points, it can be analyzed that bituminous coal exists in the form of 2, 3 rings, and was dominated by 2 rings because its X_{BP} was closer to the aromatic compound naphthalene; anthracite coal also exists in the form of 2, 3 rings but its X_{BP} was closer to the aromatic compound phenanthrene, and so is dominated by 3 rings.

2.2.3 *Characterization of Molecular Aromatic Structural Units of Coal*

Combined with peak fitting parameters of ^{13}C-NMR carbon spectrum of each coal sample in Table 2.4, the function model established by SOLUM can calculate the average carbon atom number *C* of each aromatic layer in the coal sample. The model function is shown in (Eq. 2.1):

$$X_{BP} = \frac{1 - \tanh\left(\frac{C-C_0}{m}\right)}{2} X_{BP'} + \frac{1 + \tanh\left(\frac{C-C_0}{m}\right)}{2} X''_{BP} \tag{2.1}$$

In the formula, $X_{BP'} = 1/2 - 3/C; X''_{BP} = 1 - \sqrt{6}/\sqrt{C}$, $m = 4.15; C_0 = 19.75$; the average number of bridge bonds and side chains contained in each aromatic layer is shown in (Eq. 2.2), where P_0 is the percentage of bridge bonds, as shown in (Eq. 2.3):

$$\sigma + 1 = \frac{(f_a^P + f_a^s)/C}{f_{a'}} \tag{2.2}$$

$$P_0 = \frac{f_a^P + f_a^s - f_{al'}}{f_a^P + f_a^s} \tag{2.3}$$

Two crystal structure parameters in molecular structure fragrant piece of bridge, the number of rings B.L. and number of side chains S.C. As shown in (Eq. 2.4) and (2.5) respectively:

$$\text{B.L.} = P_0(1 + \sigma) \tag{2.4}$$

$$\text{S.C.} = (\sigma + 1) - \text{B.L.} \tag{2.5}$$

Taking into account the structural parameters characterized by NMR experiments, the average molecular mass *MW* of the aromatic lamellae as well as the average molecular mass M_σ of the side chains or 1/2 bridging chains can be expressed by (Eq. 2.6) and (Eq. 2.7), respectively:

Table 2.6 Fitting parameters of coal molecular structure

Coal samples	C	$\sigma + 1$	P_0	B.L.	S.C.	MW	M_σ
Bituminous coal	15	1.71	0.3245	0.554	1.172	619.36	201.60
Anthracite	18	1.03	0.1842	0.189	0.840	592.78	196.39

$$MW = \frac{C12.01}{f_a{}'W_c/100} \tag{2.6}$$

$$M_\sigma = \frac{MW - 13M_G - 12(1 - M_G)C}{\sigma + 1} \tag{2.7}$$

where: W_c, content of C element; M_G, ratio of protonated aromatic carbon to aromatic carbon.

Based on the calculation formulas mentioned above, the fitting structural parameters of the selected two coal molecules are obtained, as shown in Table 2.6.

According to the fitting parameters of coal molecular structure (as shown in Table 2.6), the average carbon atom number of aromatic sheets of bituminous coal and anthracite is 15 and 18 respectively, and P_0 of the two coal samples is exceptionally different. The bridge bond P_0 of bituminous coal is the largest, and more bridge bonds mean that the degree of molecular condensation of coal is low, which is a characteristic of low-rank coal. The bridge bond ratio P_0 of anthracite is only 0.1842, indicating that the aromatic ring in anthracite is more directly connected to the aromatic ring than the bridge bond, which means that the aromatic group in anthracite has a higher degree of condensation.

2.3 Molecular Surface Characteristics of Coal

2.3.1 *Fourier Infrared Spectroscopy Instrument and Parameter Setting*

The combustion process of coal is essentially a process of massive heat release by coal-oxygen complex reaction. Therefore, in-depth study and understanding of the types and content of microscopic groups in coal is of vital significance for mastering the reaction characteristics of coal. Fourier transform infrared spectroscopy (FT-IR) is one of the commonly used techniques to explore the structure and properties of materials, and its advantage is that it can provide key information about chemical structure.

The infrared spectrum experimental testing instrument used in this paper is Nicolet™ iZ10 FTIR, the specific device is shown in Fig. 2.4. The changes of surface functional groups, fatty side chain structure, aromatic structure, hydroxyl group and other active groups in coal can be quantitatively analyzed by infrared spectrum test.

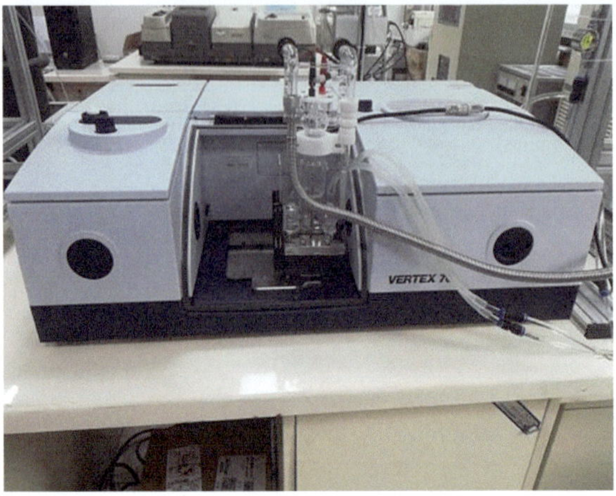

Fig. 2.4 NicoletTM iZ10 FTIR

This experiment adopted the potassium bromide tablet method as the test method (Dai et al. 2023), in which the mass ratio of potassium bromide to sample was set at 100:1 to ensure the accuracy and reliability of the test results. According to the Beer-Lambert law, in order to compare the infrared spectra of different coal samples, it is necessary to ensure that each coal sample has the same sample concentration, that is, the coal sample quality. After comprehensive consideration, the quality of coal sample selected by the experiment is more suitable in the range of 0.8–1.5 mg. Therefore, it is necessary to strictly control the mass of a single sample to about 1.0 mg during the experiment. At the same time, in order to acquire the infrared spectral information of coal samples comprehensively, the wavenumber range of the test scan is set to 400–4000 cm^{-1} to cover the vibration frequencies of various microscopic groups in coal.

2.3.2 Analysis of the Content of Reactive Groups in Coal Molecules

Coal, as a complex mixture of macromolecules, contains a wide variety of microscopic groups within its structure, such as benzene rings, thick rings, aliphatic carbon chains, ether bonds, hydroxyl and carbonyl groups (Ma et al. 2023). The infrared spectral interval can be divided into the near-infrared region (2800–4000 cm^{-1}), the mid-infrared region (400–4000 cm^{-1}), and the far-infrared region (25–1000 cm^{-1}) according to the wavelengths; the mid-infrared region is the sensitive region for common chemical groups and is the most widely used. The mid-infrared region is divided into four regions (Wang 2007; Zhang et al. 2023): (1) <1650 cm^{-1} is attributed to the X–Y stretching vibration and X–H deformation vibration region, which mainly includes the skeleton vibration of the C–C single bond, the deformation vibration of the C–H, N–H, and so on; (2) 1200–1900 cm^{-1} is attributed to

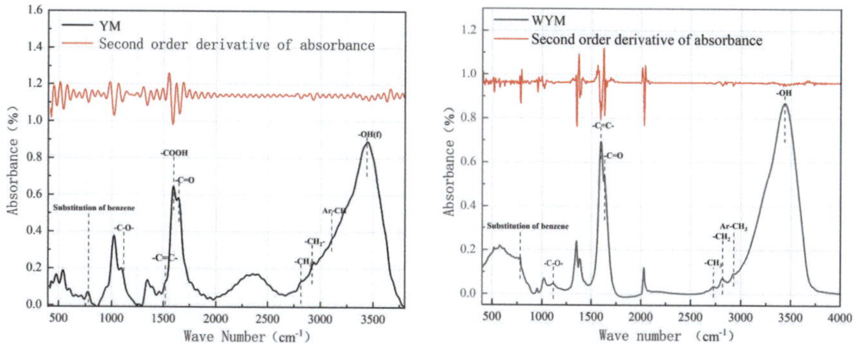

(a) Infrared spectra of bituminous **(b) Infrared spectra of anthracite coal**

Fig. 2.5 Infrared spectra of each coal sample

the double-bonding stretching vibration region, which mainly includes the skeleton vibration of the aryl ring and the vibration of C=C, C=O, C=N, $-NO_2$, aryl heterocycles and other groups of stretching vibration, etc.; (3) 2500–1900 cm^{-1} is attributed to the triple bond and cumulative double bond region, mainly including alkyne bond –C≡C–, nitrile group –C≡N, propylene dienophile group –C=C=C–, alkenonyl group –C=C=O, isocyanate group –N=C=O, etc., and other asymmetrical stretching vibration; (4) 2500–4000 cm^{-1} is attributed to the X–H stretching vibration region, mainly including the stretching vibration of O–H, N–H, C–H and S–H bonds. The Fourier infrared absorption spectra of bituminous coal and anthracite were plotted according to the experimental results, as shown in Fig. 2.5 (Table 2.7).

According to Beer-Lambert's law, the area of the absorption peaks in the infrared spectrum can characterize the content of components in the sample (Delgado 2022). Therefore, the area size of the absorption peaks in the IR spectra was used to indicate the content of their corresponding functional groups. The peak fitting process of IR spectra was mainly performed by using Peakfit software.

After deducting the baseline using the zonal deduction method, the second-order derivative method in the Peakfit software was used for automatic fitting, which is an accurate and effective method to analyze the spectral data of coal samples. Therefore, the Lorentz/Gaussian area integration method was used to calculate the area of the peaks during the fitting process of the plots, which can more accurately depict the shapes and intensities of the peaks and thus improve the accuracy of the analysis.

(a) **Analysis of Hydroxyl (–OH) Content**

Figure 2.6 shows the peak fitting diagrams of hydroxyl functional groups in two kinds of coal samples, from which it can be seen that the hydroxyl group exists in the molecular structure of coal in the form of –OH (alcohol), –OH (phenol), and –OH (bonding), and the spectral peaks attributed to them are located in the range of 3200–3700 cm^{-1}, respectively. Between 3500 and 3600 cm^{-1}, the group attributes are attributed to polyol; the peak is 3400 cm^{-1}, the group attributes are attributed to hydrogen bonding, mostly water; between 3330 and 3400 cm^{-1}, the

Table 2.7 Infrared spectral attribution of active groups

Group type		Absorption frequency/ cm^{-1}	Group properties
Hydroxyl group	–OH	3700–3650	Metal hydroxide O–H stretching, strong and sharp
		3600–3500	Polyol
		3400	Hydrogen bonded, mostly water
		3400–3330	Alcohol hydroxyl stretching vibration
		3330–3240	Phenol hydroxyl telescopic vibration
Carbon and oxygen group	C–O–C	1150–1000	C–O stretching vibration band
		1125–1110	C–O–C antisymmetric and symmetric telescopic vibrations
	C=O	1780–1765	Aromatic esters, anhydrides, and C=O bonds in peroxides
		1770–1720	Expansion and contraction vibrations of aliphatic anhydrides
		1736–1722	Carbonyl groups in aldehydes, ketones, and esters
		1690–1650	Stretching vibrations of the quinone group
	COO	1715–1690, 1720	Stretching vibrations of unsaturated carboxylic acids
		1725–1700	Stretching vibrations of saturated carboxylic acids
		1590–1560	Anti-symmetric telescopic vibration
		1420–1390	Symmetric telescopic vibration, weak
Aliphatic compound	$–CH_3$	2975–2950	Antisymmetric stretching vibration of $–CH_3$ in aliphatic or cycloalkane hydrocarbons
		2935–2915	$–CH_3$ anti-symmetric telescopic vibration

(continued)

Table 2.7 (continued)

Group type		Absorption frequency/ cm^{-1}	Group properties
		2880–2860	Symmetric stretching vibration of $-CH_3$ in aliphatic or cycloalkane hydrocarbons
		1435–1465	Antisymmetric deformation vibration
		1370–1380	Symmetrical deformation vibration
	$-CH_2-$	2850–2840	Symmetrical telescoping vibration
		1460–1470	Symmetrical bending vibration
		1439–1449	Bending vibration
	–CH	2890–2900	C–H telescopic vibration
	$=CH_2$	3080	Against telescoping
		3000	Symmetrical expansion and contraction
	=C–H	3040	Telescoping vibration
	–C=C–	1560–1460	Aromatic Ring Skeleton
		1635–1595(1600)	Expansion and contraction vibrations of –C=C–
		1620–1680	
Aromatic compound	–CH	3100–3000	Aromatic hydrocarbons –CH
	Substitution of benzene	900–850	Pentasubstituent on benzene ring
		860–800	Benzene ring p-disubstituted
		800–750	Disubstitution between benzene rings
		770–730	Benzene ring o-disubstituted
		710–690	Disubstitution between benzene rings

group is attributed to alcohol hydroxyl stretching vibration; and between 3240 and 3330 cm^{-1}, the group is attributed to phenol hydroxyl stretching vibration.

Figure 2.7 shows the percentage of hydroxyl group content in the characteristic spectrum of hydroxyl peaks. As can be seen from the figure, the percentage of

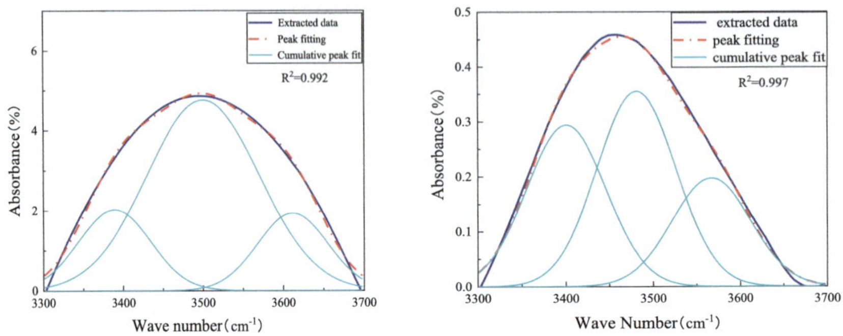

(a) Bituminous coal hydroxyl fraction fitting **(b) Anthracite coal hydroxyl fraction fitting**

Fig. 2.6 Fitting of hydroxyl reactive group peak splitting

alcohol-phenol group content in bituminous coal samples is exceedingly close to 3.6 and 4.08%, respectively; the percentage of alcohol group content in anthracite samples is the lowest, which is only 1.6%; and the total hydroxyl group content in the three types of anthracite and bituminous coal is 12.19 and 18.16%, respectively, which is obvious that the total hydroxyl group content in the bituminous coal is higher. From the different hydroxyl groups in the figure, it can be seen that the hydroxyl groups contained in bituminous coal samples are mainly dominated by –OH.

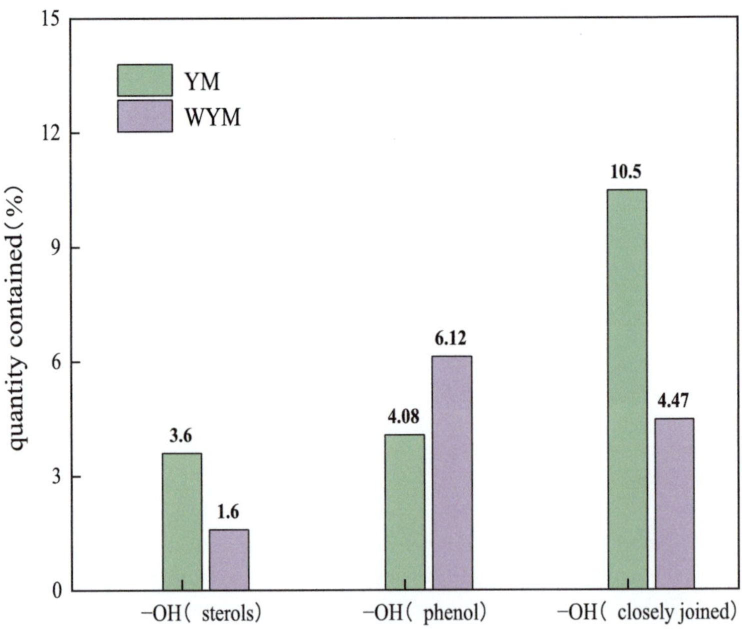

Fig. 2.7 Hydroxyl characteristic spectral peak group contents

(b) Carbon–oxygen Group Content Analysis

Figure 2.8 shows the peak fitting diagram of carbonyl groups, from which it can be seen that the spectral bands of carbonyl groups in coal are mainly distributed between 1000–1300 and 1600–1800 cm^{-1}, among which the carbonyl group (–C=O–) contains all kinds of ketones, aldehydes, and ester groups, with peaks located between 1722 and 1780 cm^{-1}, and the carboxyl group (–COO) is located between 1420 and 1720 cm^{-1}. The contents of both carbonyl and carboxyl groups showed a prominent decreasing trend with the increase of coal deterioration degree, while the contents of –C–O– group showed a substantial increasing trend with the increase of coal deterioration degree. The carboxyl group in anthracite is so small that the absorption peaks in the spectrum are merged into the carbonyl group on its side, which makes the proportion of –C–O– group in anthracite slightly higher than that of bituminous coal, and the large reduction of carboxyl group means the decrease of hydroxyl group, which is extremely active in the process of spontaneous combustion of coal, which indirectly reflects that the spontaneous combustion tendency of the coals with high degree of metamorphism is lower.

Figure 2.9 shows the percentage of carbon and oxygen group content. From the figure, it can be seen that the carbon and oxygen groups of bituminous coal and anthracite coal samples are mainly dominated by –C–O–, and the percentage of their group content is exceptionally close to each other, which is 20.2 and 22.2%, respectively; the distribution of carbon and oxygen groups in the anthracite coal has a remarkable feature, and the carboxyl group content of anthracite coal in the two raw coal samples is extremely low. The organic substances in the coal undergo a condensation reaction, which leads to the decomposition or transformation of the carboxyl groups and other functional groups in large quantities, thus reducing their content.

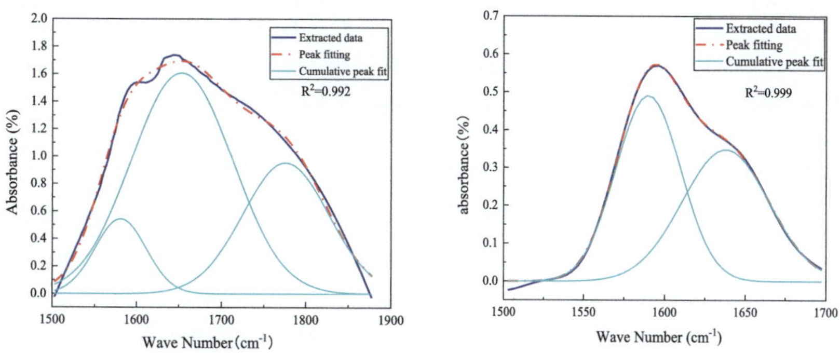

(a) Bituminous coal carbon-oxygen group split-peak fit

(b) Anthracite coal carbon-oxygen group split-peak fit

Fig. 2.8 Carbon–oxygen group split peak fitting

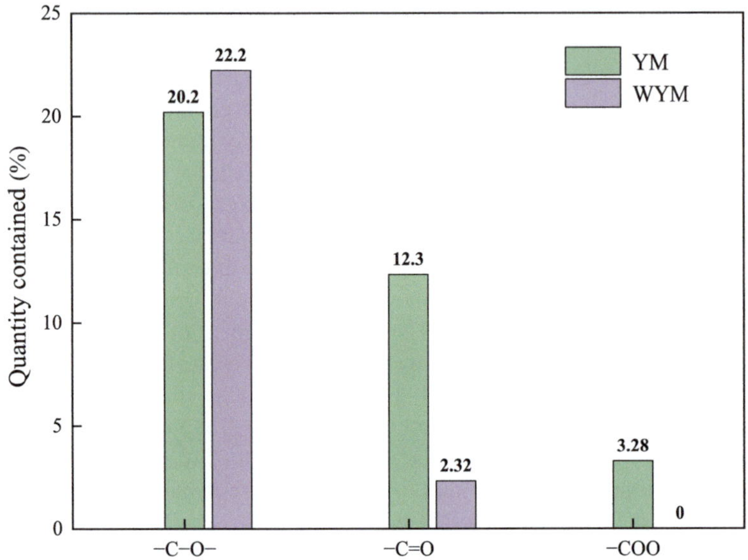

Fig. 2.9 Characteristic spectral peak content of carbon and oxygen groups

(c) Analysis of Aliphatic Hydrocarbon Content

Figure 2.10 shows the peak fitting diagram of aliphatic functional groups, from which it can be seen that the spectral bands of aliphatic hydrocarbons are mainly located in 2800–3000 cm^{-1}, of which 2950–2975 cm^{-1} peaks are attributed to the $–CH_3$ antisymmetric stretching vibration of aliphatic hydrocarbons or cycloalkanes, 2915–2935 cm^{-1} belongs to the $–CH_3$ antisymmetric stretching vibration, and the peaks of 2840–2850 cm^{-1} are attributed to the methylene ($–CH_2–$) symmetric stretching vibration. The peaks at 2840–2850 cm^{-1} were attributed to the symmetric stretching vibration of methylene ($–CH_2–$). Methyl and methylene were detected in all the experimental coal samples, and the methyl and methylene absorption intensities were higher overall, indicating that there were more side chains in the molecular structure, but there were some differences in the methylene absorption intensities of different coal samples.

Figure 2.11 shows the content ratio of aliphatic hydrocarbon characteristic spectral peak groups, and the results of peak splitting from the figure demonstrate that the aliphatic functional groups of the two coal samples are mainly dominated by methyl ($–CH_3$), and the content of methyl in bituminous coal and anthracite was 8.2 and 12.3%, respectively, which was higher than that of methylene, and also the content of methylene is the highest among the two types of coal samples in anthracite, which is 9.24%. By comparing the aliphatic carbon functional group contents of the two coal samples, it was found that the absorption content of methylene and methyl in anthracite with a higher degree of metamorphism accounted for extremely higher than that of bituminous coal with a lower degree of metamorphism. The high branching

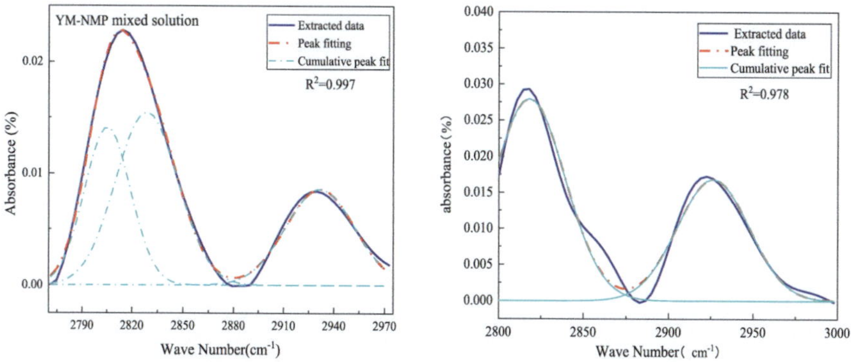

(a) Bituminous coal aliphatic peak fitting **(b) Anthracite aliphatic peak fitting**

Fig. 2.10 Aliphatic reactive group peak fitting

ratio means that there are far fewer long chains than short chains in the coal molecules, and the long-chain molecules are more likely to break than the short-chain molecules when absorbing the same amount of energy, and then the reaction occurs, in which the two coal samples selected, the anthracite coal has a higher molecular stability.

(d) Aromatic Hydrocarbon Content Analysis

The degree of metamorphism of the coal species determines the proportion of aromatic hydrocarbon structures in the coal molecule. The spectral bands of aromatic

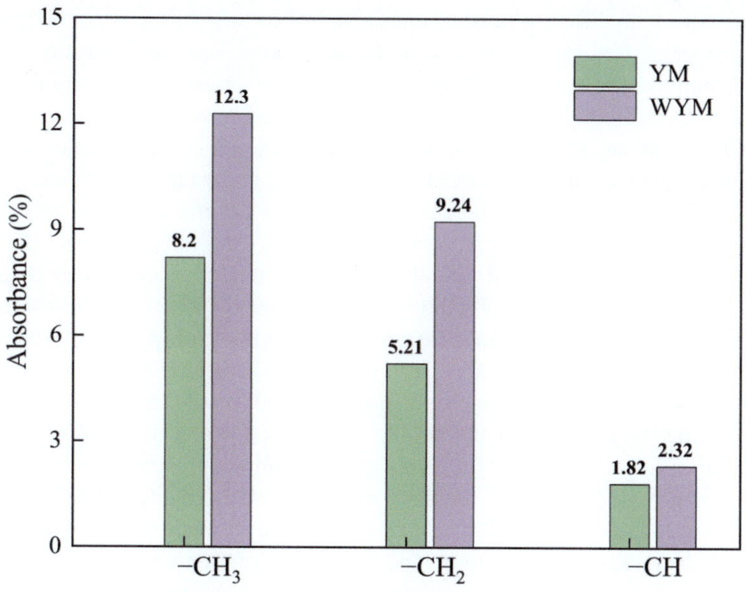

Fig. 2.11 Aliphatic hydrocarbon characteristic spectrum peak group content

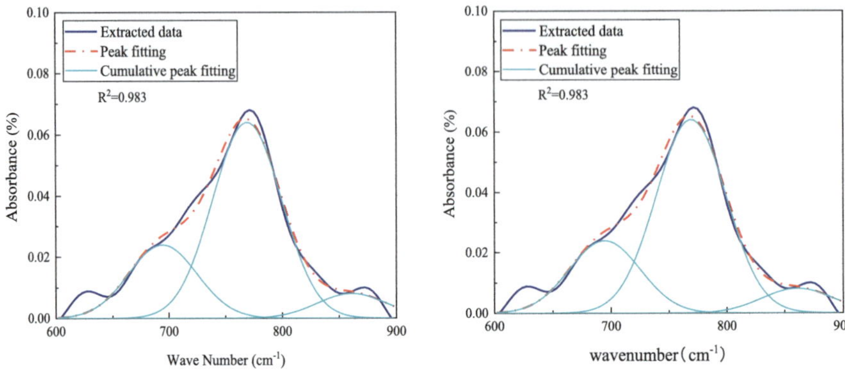

(a) Bituminous coal aromatic peak fitting **(b) Anthracite aromatic peak fitting**

Fig. 2.12 Aromatic reactive group peak fitting

hydrocarbons are distributed in different regions, and the specific peak fitting results are shown in Fig. 2.12. Between 700 and 900 cm^{-1}, the group properties are attributed to the deformation vibration of multiple-substituted aromatic hydrocarbons, Ar–CH, and there are several obvious peaks, which represent different substitution types of aromatic hydrocarbons; between 1460 and 1560 cm^{-1}, the group properties are attributed to the aromatic ring skeleton; between 1600 and 1635 cm^{-1}, the group properties are attributed to the stretching vibration of –C=C–; and between 3000 and 3100 cm^{-1}, the group properties are attributed to the aromatic hydrocarbon –CH. The main aromatic hydrocarbons of the two samples are attributed to the aromatic hydrocarbons with different substitutions, and the positions of the peaks of the functional groups are tremendous obvious in the cumulative peaks of the aromatic group in the fitting diagrams. The peak positions of the functional groups are distinctly obvious in the aromatic accumulation peak fitting diagrams.

The results of the percentage of aromatic hydrocarbon group content of the two coal samples with aromatic hydrocarbon containing groups at 700–900 cm^{-1} peak fitting are shown in Fig. 2.13. The total aromatic (–C=C–, aromatic –CH, substituted benzene) contents of bituminous coal and anthracite coal samples were 26.84 and 38.64%, indicating that the total aromatic content of anthracite coal with a high degree of metamorphism was higher. The absorption peak group content of these two coal samples in the range of 1600 cm^{-1} is larger than that in the range of 900–700 cm^{-1}, which is mainly because the –C=C– double bond characterized by 1600 cm^{-1} is the main structure of the aromatic ring, and the original coal samples have not been oxidized, and the structure of the double bond is stable and the content is larger, so it shows a larger intensity of the spectral peak. In addition, the content of each aromatic group contained in anthracite coal varies greatly, and the –C=C– group of anthracite coal has the largest content of 31.2%, while aromatic –CH and substituted benzene have relatively less content, and the content of the two groups was 5.2 and 2.24%, respectively. The content of aromatic –CH of bituminous coal was lower, only 2.33%. According to the content of aromatic hydrocarbon groups, it can be

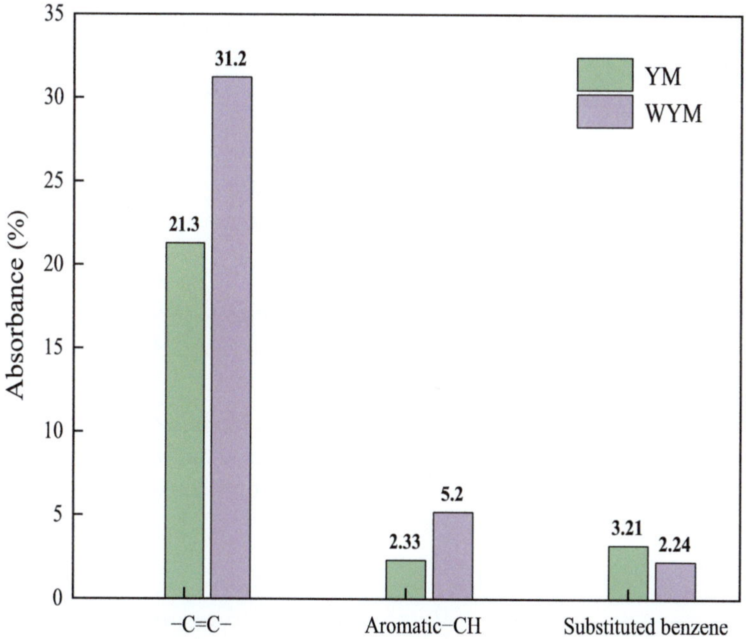

Fig. 2.13 Aromatic hydrocarbon characteristic spectral peak group content

seen that bituminous coal has more substituent hydrocarbons than anthracite, and the aromatic hydrocarbons appear more in the form of substituent aromatic rings.

From the figure, it can be seen that the content of different functional groups in bituminous coal and anthracite coal samples has a large difference in the content of each functional group is summarized in Table 2.8. After comparing the content of functional groups in the two types of coal samples, it was found that the content of oxygen-containing groups in bituminous coal was the largest in the proportion of the functional groups in each region, which was 35.8%, but the content of aliphatic hydrocarbon groups in the functional groups in the region is the smallest, which was 15.2%; the content of aromatic hydrocarbons in anthracite coal. The content of aromatic hydrocarbon functional group in anthracite coal was the largest proportion of functional groups in each region, which was 38.7%, but the content of hydroxyl functional group is the smallest proportion in the region, which is only 9.2%. As indicated by the data, the absorption content of methylene ($–CH_2–$) methyl ($–CH_3$) in anthracite with higher degree of metamorphism is significantly higher than that in bituminous coal with lower degree of metamorphism. However, as the organic substances in anthracite undergo polycondensation reactions, this process results in a large number of decomposition or transformation of functional groups such as carboxyl groups, resulting in a exceedingly low content of carboxyl (–COO) groups in anthracite.

Table 2.8 Proportion of reactive groups in each region of coal samples with different degrees of deterioration

Coal	Proportion of reactive groups in each region of the coal sample (%)			
	Aromatic hydrocarbon	Carbon–oxygen group	Aliphatic hydrocarbon	Hydroxyl group –OH
Bituminous coal	26.816	35.828	15.234	17.152
Anthracite coal	38.655	24.552	23.85	9.185

2.3.3 Characterization of Coal Molecular Infrared Parameters

Utilization of apparent aromaticity of coal samples in infrared spectroscopy f_a characterizing aromatic carbon. The carbon atoms in the coal molecule are organized as aromatic carbon (C_{ar}) and aliphatic hydrocarbon carbon (C_{al}) two forms exist. f_a can be expressed by the Eqs. (2.8)–(2.10):

$$\frac{H_{al}}{H} = \frac{H_{al}}{H_{al} + H_{ar}} = \frac{A_{2800-3000}}{A_{2800-3000} + A_{700-900}} \tag{2.8}$$

$$\frac{C_{al}}{C} = \left(\frac{H_{al}}{H} \times \frac{H}{C}\right) \Big/ \frac{H_{al}}{C_{al}} \tag{2.9}$$

$$f_a = 1 - \frac{C_{al}}{C} \tag{2.10}$$

Formulas, H_{al}/H is ratio of aliphatic (T_0) concentration to hydrogen atom (T_0), H_{al} depends on the area of the integrated absorption band where the absorption peak is located at 2800–3000 cm^{-1}; depends on the area of the integrated absorption band where the absorption peak is located at 700–900 cm^{-1}. Depends on where the absorption peak is located 2800–3000 cm^{-1} area of the integrated absorption zone; H_{al} depends on the area of the integrated absorption band where the absorption peak is located at 700–900 cm^{-1}.

C_{al}/C, aliphatic carbon fraction; H/C, ratio of hydrogen atoms to carbon atoms; H_{al}/C_{al}, ratio of hydrogen to carbon in the aliphatic group, taking the value of 1.8; *I* is another indicator of aromaticity and coal rank and is used to compare the relative abundance of aromatic and aliphatic functional groups. *I* is represented by (Eq. 2.11):

$$I = \frac{A_{700-900}}{A_{2800-3000}} \tag{2.11}$$

formulas containing, $A_{2800-3000}$, occupied by spectral peaks located at 2800–3000 cm^{-1}, was used to estimate the total aliphatic CH ($-CH_3$, $-CH_2-$, CH) quantity contained.

Table 2.9 Structural parameters determined by FTIR spectroscopy of coal samples

Coal sample	f_a	I	DOC	A ($-CH_2-$)/ A($-CH_3$)	'*C*'
Bituminous coal	0.5737339	0.2758278	1.1036357	0.6361307	0.6683151
Anthracite coal	0.7067709	0.2817779	1.7121076	0.5071053	0.4608037

The intensity of aromatic C=C is related to the area (A_{1600}) under the peak of the 1600 cm^{-1} spectrum, then the degree of condensation (*DOC*) of the aromatic ring can be determined:

$$DOC = \frac{A_{700-900}}{A_{1600}} \tag{2.12}$$

The length of the aliphatic chain and the extent of branched aliphatic side chains were calculated from the $-CH_2-/-CH_3$ intensity ratio, A($-CH_2-$)/A($-CH_3$) Higher values of A($-CH_2-$)/A($-CH_3$) correspond to longer aromatic ring aliphatic chains, whereas lower values indicate a relatively compact structure with less space between aromatic rings. The $-CH_2-/-CH_3$ intensity ratio is estimated:

$$\frac{A(-CH_2-)}{A(-CH_3)} = \frac{A_{2915-2940(2922)}}{A_{2950-2975(2952)}} \tag{2.13}$$

Formula, $A_{2915-2940(2922)}$ and $A_{2950-2975(2952)}$ the asymmetric tensile strength depends on the $-CH_2-$ and $-CH_3$ groups, respectively.

The structural parameter '*C*' of oxygen-containing groups represents the change in the ratio of C=O to C=C groups to characterize the maturity of the coal. '*C*' depends on (Eq. 2.14).

$$'C' = A_{1650-1800}/A_{1650-1800} + A_{1600} \tag{2.14}$$

To further investigate the molecular structure of bituminous coal and anthracite, the infrared spectral data of the coal samples were used to calculate each structural parameter of coal using (Eq. 2.11–2.14), and the specific results are shown in Table 2.9.

As can be seen from Table 2.9, the ratio of A($-CH_2-$)/A($-CH_3$) reflects the length of the fatty chains and the degree of branched fatty side chains in the coal, in which the A(CH_2-)/A($-CH_3$) value of anthracite coal is lower, about 0.5071, indicating that the fatty chains are usually in the form of shorter chains in anthracite coals with a much higher degree of metamorphism.

References

Bhatta R, Dash N (2023) Influence of particle size on maceral and mineral distribution in lower Gondwana coals of India. Arab J Geosci 16(3):188. https://doi.org/10.1007/s12517-023-11286-4

Chang T, Zhou X, Deng X et al (2023) Correction of sideband effects of nuclear magnetic resonance carbon spectrum in coal and its application in coal structure analysis. Energy Sci Eng 11(8):2763–2774. https://doi.org/10.1002/ese3.1490

Dai F, Zhuang Q, Huang G et al (2023) Infrared spectrum characteristics and quantification of OH groups in coal. ACS Omega 8(19):17064–17076. https://doi.org/10.1021/acsomega.3c01336

Delgado R (2022) Misuse of Beer-Lambert Law and other calibration curves. Royal Soc Open Sci 9(2):211103. https://doi.org/10.1098/rsos.211103

Li HT, Zhu ZR, Qiao JW et al (2022) Molecular representations of Jurassic-aged vitrinite-rich and inertinite-rich coals in Northern Shannxi province by FTIR, XPS and ~(13)C NMR. Spectrosc Spect Anal 42(08):2624–2630. https://doi.org/10.3964/j.issn.1000-0593(2022)08-2624-07

Ma T, Zhai X, Xiao Y et al (2023) Study on the influence of key active groups on gas products in spontaneous combustion of coal. Fuel 344:128020. https://doi.org/10.1016/j.fuel.2023.128020

Qin HL (2019) Macromolecular structure of maceral composition in Hancheng Mine Area, Shanxi

Wang XF (2007) Infrared spectral research on the coal oxidized spontaneous combustion process. Liaoning

Xiang JH, Zeng FQ, Liang HZ et al (2016) Carbon structure characteristics and evolution mechanism of different rank coals. J China Coal Soc 41(6):1498–1506

Zhang W, Wang L, Su LH (2023) Carbon structure characteristics and evolution mechanism of different rank coals. Inner Mongolia Coal Econ 16:10–12

Zhang JG (2015) Ash fusion behaviors and transformation of coal minerals during coal conversion process, Chongqing

Zhu B, Dong X, Fan Y et al (2023) Structural characterization and molecular model construction of high-ash coal from northern China. Molecules 28(14):5593. https://doi.org/10.3390/molecules28145593

Chapter 3
Modelling of Coal Auto-Ignition Reactive Groups and Their Weak Interactions with Extractants

Abstract Based on the skeleton structure and surface group characteristics of coal samples, molecular model of the key active groups for coal spontaneous combustion were constructed in this chapter. Based on density functional theory (DFT), the electrostatic potential distribution and active center of reaction were simulated and calculated. Then, importance of reactive groups such as hydroxyl, carbonyl and methyl groups in the coal spontaneous combustion were emphasized. Hence, active sites of coal molecules could be more accurately predicted. Furthermore, weak interaction mechanisms between the molecular structure of the extractant and the active group molecules were analyzed by optimizing the extractants.

3.1 Molecular Modelling of Coal Auto-Ignition Reactive Groups and Their Electrostatic Potential Distribution Characteristics

All the non-periodic calculations in this chapter are based on DFT, and Gaussian 16 software was used to optimise the calculations for the established coal sample model and the main components of the extractant. The calculation method is B3LYP hybridisation flooding method, and the calculation basis group is 6-311G(d) (Frisch et al. 2013), and the frequency analysis was carried out with this calculation method and calculation basis group, and the optimal molecular structure was obtained after the optimisation of the structure frequency was completed. The dispersion correction was further calculated using Grimme's D3 (BJ) optimisation method (Grimme et al. 2010; Gonzalez and Schlegel 1989). Meanwhile, the stability of the Kohn–Sham determinant wavefunction in DFT was tested. If an unstable wave function is found (Xu et al. 2023), the limits will be appropriately lowered, the optimisation will be re-done, and the stability will be tested again until a stable wave function is found, and finally the electrostatic potential of the optimised molecular structure is calculated and plotted using Multiwfn and VMD software (Wang 2023). The calculation parameters were set as follows: Job Type: Opt + Freq; Method: Ground State, DFT, Default Spin, B3LYP; Basis set: 6-311G (d).

J. Deng et al., *Coal Spontaneous Combustion Theory with Chain Self-promoted Oxidizing Induced by Active Group*, https://doi.org/10.1007/978-981-96-5362-1_3

3.1.1 Molecular Modelling of Active Groups for Coal Auto-Ignition

The chemical structure of coal as a complex macromolecular network system has never been fully understood. The generally accepted view of the current coal structure theory is that coal molecules are formed by many similar basic structural units connected by bridge bonds (Luo et al. 2023; Ma et al. 2018). From the analyses in Chap. 2, it can be seen that the basic structural units of the YM molecule selected in this paper are mainly bicyclic, and the basic units of the WYM molecule are mainly tricyclic, and the numbers of carbon atoms on the aromatic lamellae are 15 and 18, with an aromatic ring bridge-to-carbon ratio (X_{BP}) of 0.289 and 0.337, which are closer to naphthalene and phenanthrene, respectively. There are four main groups on the surface of the molecule, which are Aliphatic side chains, –OH groups, oxygen-containing functional groups other than –OH, and aromatics. Compared to WYM, YM, owing to its lower degree of metamorphism, has mainly methylene ($-CH_2-$) as the basic unit of its aliphatic side chain, whereas the WYM side chain is dominated by methyl ($-CH_3$). Oxygen-containing groups in both coal samples mainly existed in the form of hydroxyl groups (–OH), ether bonds (–O–), carbonyl groups (–C=O–), and an exceedingly small amount of carboxylic acids (–COOH). Among them, the –OH group had the highest content in both coal samples, followed by ether bond (–O–). However, because of the more stable structure of ether bond (–O–), it is difficult to react at low temperatures, whereas carboxylic acid (–COOH) is almost undetectable in WYM. Therefore, combining the results of previous studies (Kato et al. 2016; Chen et al. 2016; Shiga et al. 2020), in this study, the dicyclic naphthalene was chosen as the core structure when constructing the molecular model of YM reactive moiety, and the hydroxyl group (–OH), carbonyl group (–C=O–) and ethyl group ($-CH_2CH_3$) substituted the H on the naphthalene ring to form the side chain; and the phenanthrene was chosen as the core structure when constructing the molecular model of WYM reactive group, with hydroxyl (–OH), carbonyl (–C=O–) and methyl (–C=O–) as the core structure, and the phenanthrene as the core structure. (–OH), carbonyl (–C=O–) and methyl ($-CH_3$) substituting H on the naphthalene ring to form the side chain. Since the ratio of substituted aromatic hydrocarbons to –C=C– in YM and WYM shows that the content of aromatic substituted hydrocarbons in both coal samples is low, in order to further determine whether the benzene ring is involved in the reaction during the low-temperature oxidation process, the benzene ring is chosen as the fourth molecular side chain in both coal samples in this chapter. The model molecular structures constructed in this paper are shown in Table 3.1.

Table 3.1 Structural models of coal molecular model

Coal species	Side-chain group	Molecular formula
YM	–OH	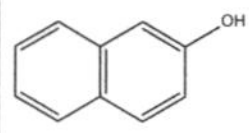
	–CHO	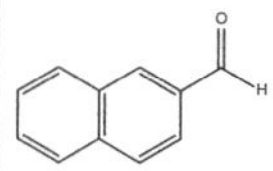
	$-CH_2CH_3$	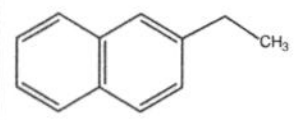
	Benzene rings	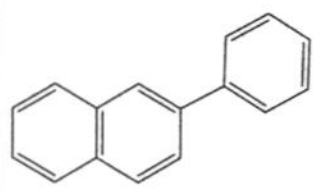
WYM	–OH	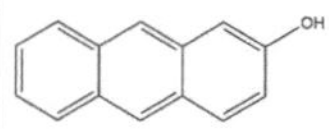
	–CHO	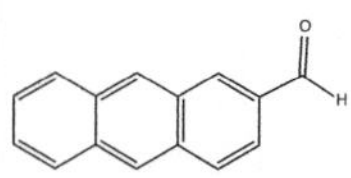
	$-CH_3$	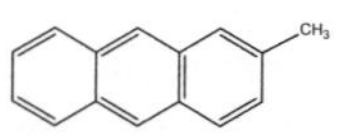
	Benzene rings	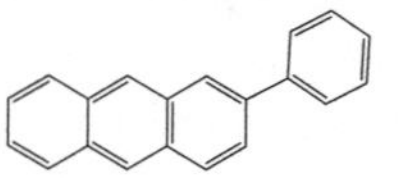

3.1.2 Characteristics of the Electrostatic Potential Distribution of YM Active Group Molecular Models

Based upon quantum chemical theory (Wang et al. 2007), we can calculate the electrostatic potential (ESP) at any position in the space around a molecule to determine its active site more precisely. In this paper, this is illustrated using two methods for describing the distribution of the electrostatic potential, the electrostatic potential surface and the electrostatic potential contour plot. The electrostatic potential surface (Sangeetha et al. 2023) is a surface map obtained by colouring each point on the surface of a molecule according to its electrostatic potential level, which can visually reflect the high or low electrostatic potential distribution on its molecular surface. In this paper, we choose to use the equivalent surface with an electron density of 0.001 a.u. to obtain the projection map of the electrostatic potential surface, and the closer the atom is to the point of maximum (minimum) electrostatic potential on the molecular van der Waals surface, the more likely it is to be a nucleophilic (electrophilic) site.

Figure 3.1 shows the electrostatic potential diagram of a double aromatic ring molecule with side chains connecting –OH, –CHO, ethyl and benzene rings, respectively. The double aromatic ring molecular model is based upon the results of NMR experiments, and exists mainly in bituminous coals with low degrees of deterioration.

In Fig. 3.1, the electrostatic potentials of the molecular surfaces of the bis-aromatic rings carrying different reactive groups are different. The lowest potentials of the four YM reactive group molecular models are all inside the aromatic ring except the electrostatic potentials at the side chains, which is caused by its special π-electron conjugated structure; the H atoms of the aromatic ring show positive potentials, and the H atoms interact with each other to form multiple regions of extreme values of electrostatic potentials. In terms of the distribution of the electrostatic potential near the side chain, the YM reactive group molecular model with a benzene ring as the side chain has an overall higher electrostatic potential at the side chain, followed by the YM reactive group molecular model with $-CH_3$. Both molecular models with different side chains show weak polarity due to the absence of O atoms with lone pair of electrons; the electron-rich region of the YM reactive group molecular model with a –CHO side chain is near the O atom of –CHO, with a lower potential, which is due to the presence of the C=O bond, resulting in the π-electrons therein being free to conjugate between the C and O atoms, which is manifested as a higher density of electrons at that place.

The positive electrostatic potential region of the coal molecule model with –OH side chain is concentrated around the aromatic ring and near the H atom on –OH, and there is a point of maximum electrostatic potential of the molecule around the H atom; the negative electrostatic potential region of the molecule is concentrated around the aromatic ring and near the H atom on –OH, and there is a point of maximum electrostatic potential of the molecule around the H atom; the negative electrostatic potential region of the molecule is concentrated around the aromatic ring and around the H atom, and there is a point of maximum electrostatic potential of the molecule; the negative electrostatic potential region of the molecule is concentrated around the aromatic ring and the H atom. The negative electrostatic potential of the molecule is concentrated in the aromatic ring and around the O atom on OH, and there is a very small value of the electrostatic potential of the molecule around the O atom. The difference in the potential around the H atom and the O atom on the –OH side chain is due to the fact that the electronegativity of the O atom is much stronger than that of the H atom, and the electrons in the covalent bond formed between the two are shifted to the O atom, which leads to the enrichment of electrons in the vicinity of the O atom on –OH, which has a lower electric potential., while the H atom presents a high potential around it due to the exposed nucleus. The difference in the side chains leads to the fact that although the potential maxima exist in all molecular models, by comparing the surface potential maxima of the molecular models with different YM reactive groups, it can be seen that the potential maxima of the molecular models with O atoms (side chains of –OH and –CHO) are higher than that of the molecular models without oxygen atoms (side chains of benzene ring and $-CH_3$), which implies

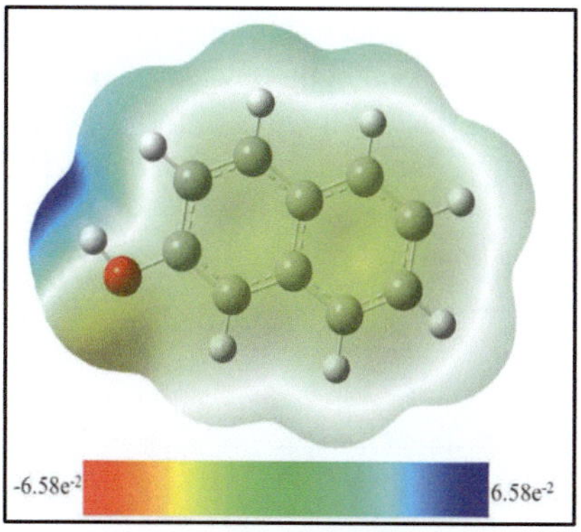

（a）**Ar-OH ESP distribution chart**

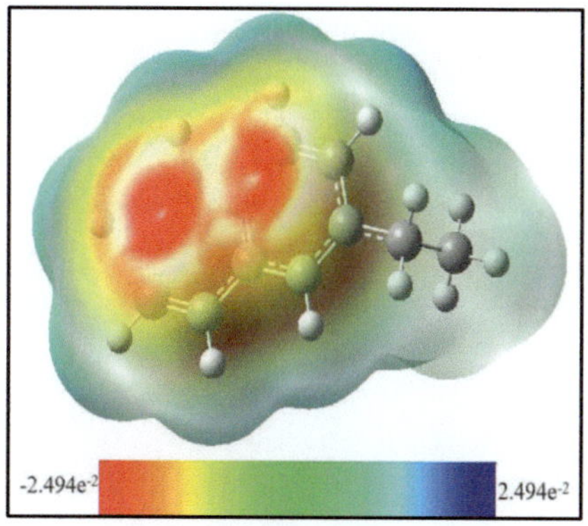

（b）**Ar-CH2CH3 ESP distribution chart**

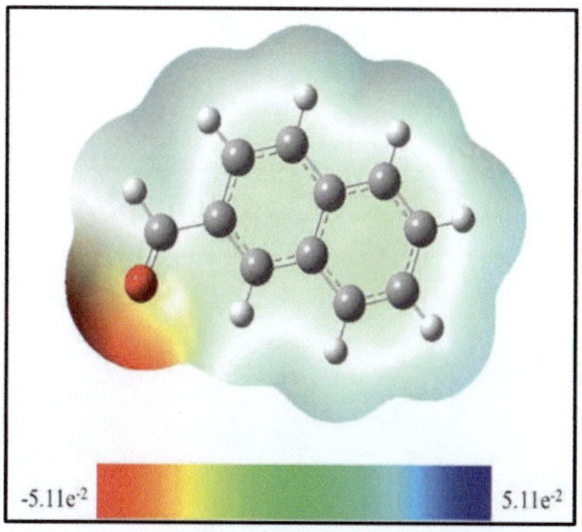

（c）**Ar-CHO ESP distribution chart**

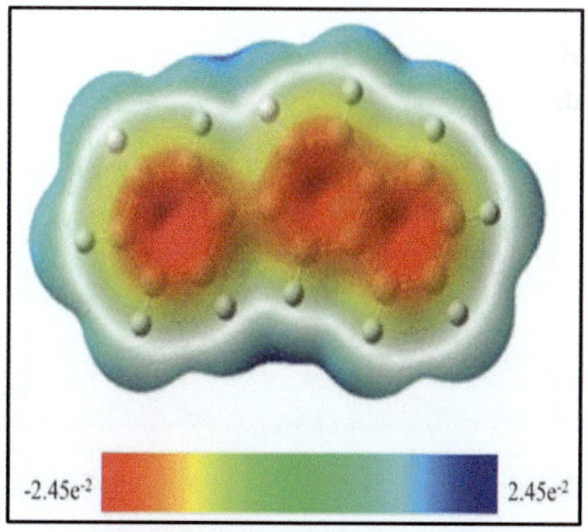

（d）**Benzene ring side chainESP distribution chart**

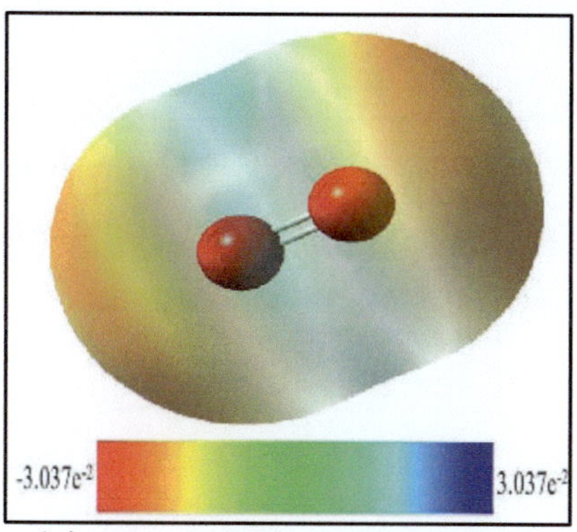

（e）**Oxygen ESP distribution chart**

Fig. 3.1 Electrostatic potential surface of the YM molecular model (with the plane in which the molecules are located as a reference) that the O atoms are more susceptible to the formation of electrophilic reaction in the coal molecule sites, and also indicates that molecules containing oxygen-containing groups are more likely to react with oxygen. The surface properties of the molecule were analysed as shown in Fig. 3.2, yielding positive potentials near all H atoms and negative potentials near all O atoms.

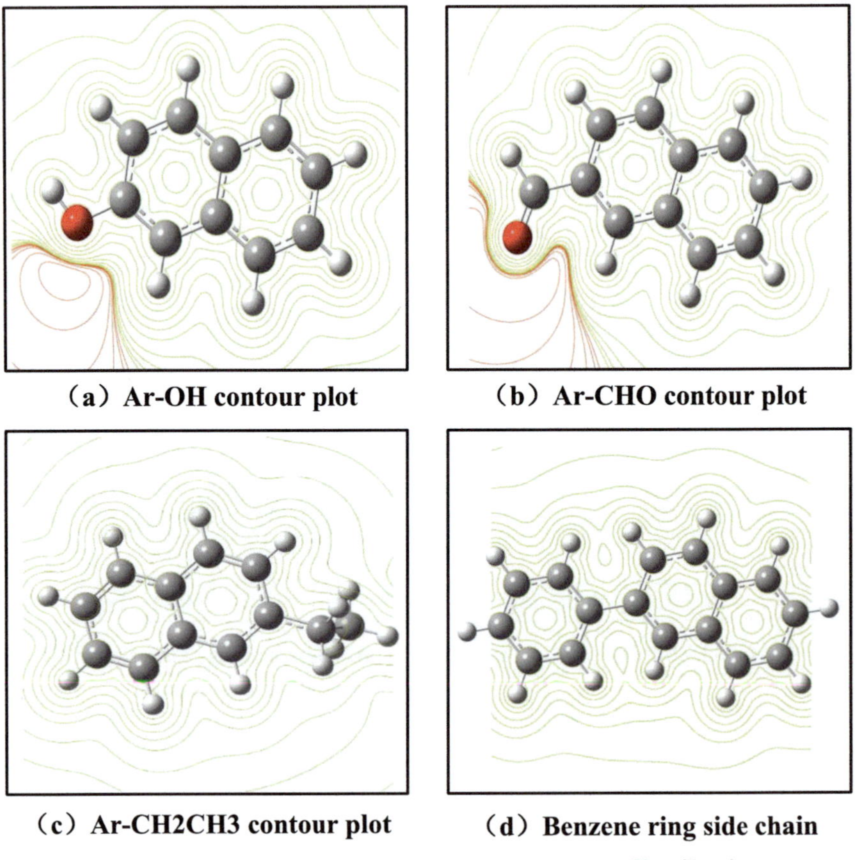

(a) Ar-OH contour plot

(b) Ar-CHO contour plot

(c) Ar-CH2CH3 contour plot

(d) Benzene ring side chain contour distribution

Fig. 3.2 Potential contour plots for the YM molecular model

3.1.3 Characteristics of the Electrostatic Potential Distribution of the WYM Reactive Group Molecular Model

Figure 3.3 shows the electrostatic potential of tri-aromatic ring molecules with side chains connecting –OH, –CHO, $-CH_3$ and benzene rings, respectively. The tri-aromatic ring molecules are modelled on the basis of NMR experiments, which are mainly present in WYMs with high degree of metamorphism.

In Fig. 3.3, the surface electrostatic potentials of the molecules with three aromatic rings carrying different reactive groups are different, but their distribution patterns are similar to those in YM molecules. By comparing the highest surface potentials of molecular models with different WYM reactive groups, it can be seen that when the molecule carries aliphatic and aromatic groups, its reactivity is lower; when the molecule carries oxygen-containing reactive groups, the positive electrostatic potentials are distributed around the periphery of the aromatic ring and around the H-atoms of the oxygen-containing reactive groups, while the negative electrostatic potentials are concentrated around the O-atoms of the oxygen-containing reactive groups, whose reactivity is higher (Fig. 3.4).

By analysing the surface electrostatic potential of the molecular models of YM and WYM reactive groups, it was calculated that the positive electrostatic potential on the molecular surface was mainly located at the pro-atoms of the bare nuclei, while the negative electrostatic potential on the molecular surface was dominated by the lone pair of electrons carried by the O atoms, and that the negative potential of the groups carrying the side chains containing the O atoms was concentrated at the O atoms on the side chains, and the negative potential was higher, which implies that the O-containing groups are more prone to be combined with the O atoms during the low-temperature oxidation stage, and the results show similarity with YM.

3.2 Extraction Agent Optimization and Electrostatic Potential Distribution Characteristics

3.2.1 Extraction Agent Optimization

The essence of extraction behavior is the formation of non-covalent bonds between organic solvents and coal macromolecules, such as hydrogen bonds, dispersion force, induction force, π–π stacking, etc., to achieve the dissolution of extracted substances by extractants (Liu et al. 2024; Çağlayan and Ipek 2021). If some models of extractant molecules and coal molecules can produce relatively stable weak interaction through the formation of non-covalent bonds, it means that the extractant has good solubility to such molecular models, that is, it has a good extraction ability.

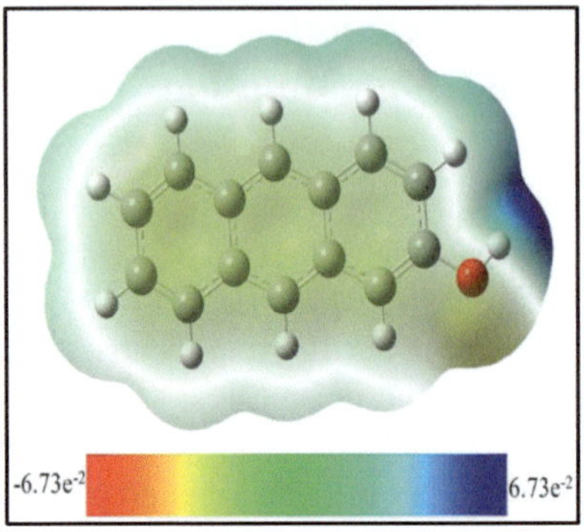

(a) Ar-OH ESP distribution diagram

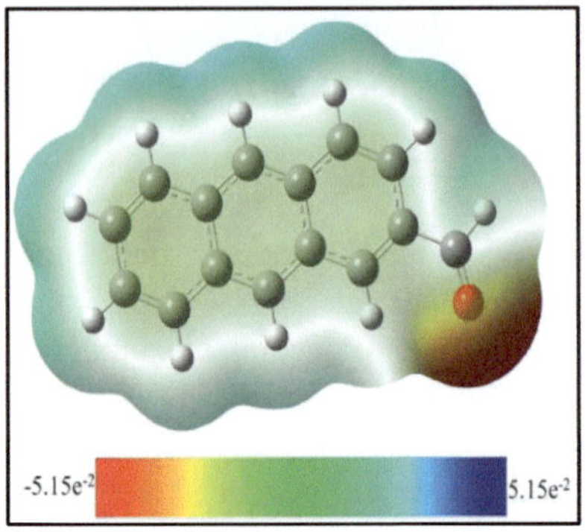

(b) Ar-CHO ESP distribution diagram

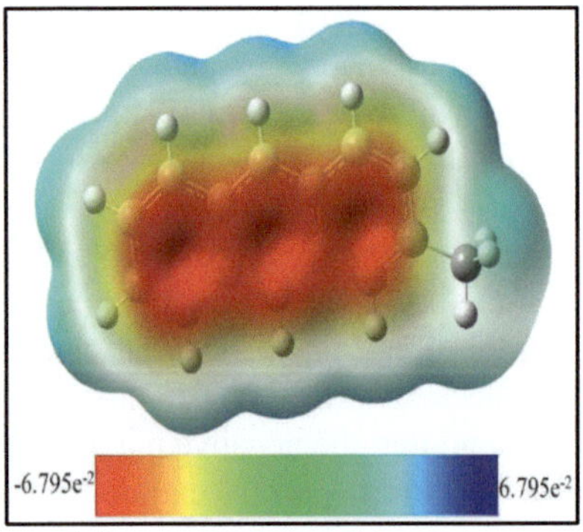

(c) Ar-CH3 ESP distribution diagram

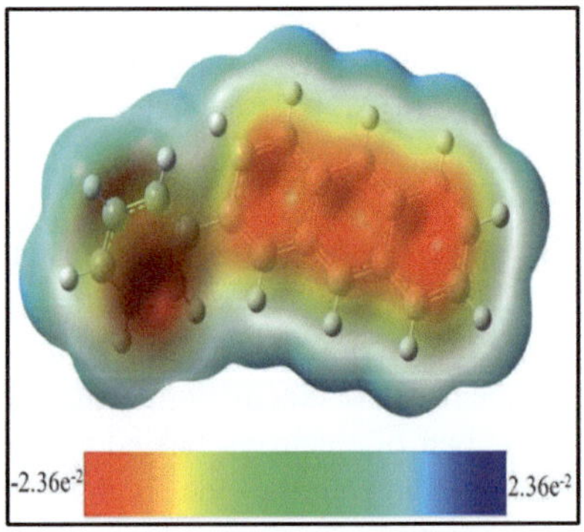

(d) ESP distribution of benzene ring side chain

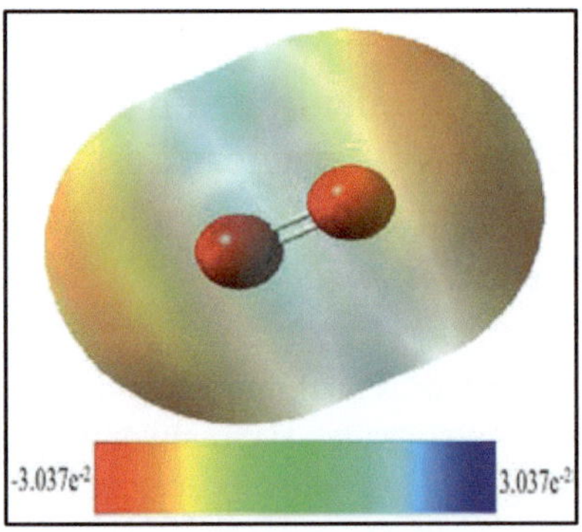

(e) Oxygen ESP distribution map

Fig. 3.3 Electrostatic potential surface of the WYM molecular model (with the plane where the molecule is located as a reference)

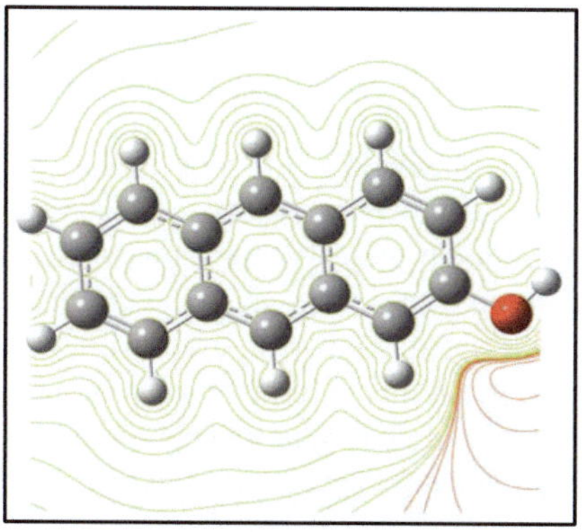

(a) Ar-OH contour distribution

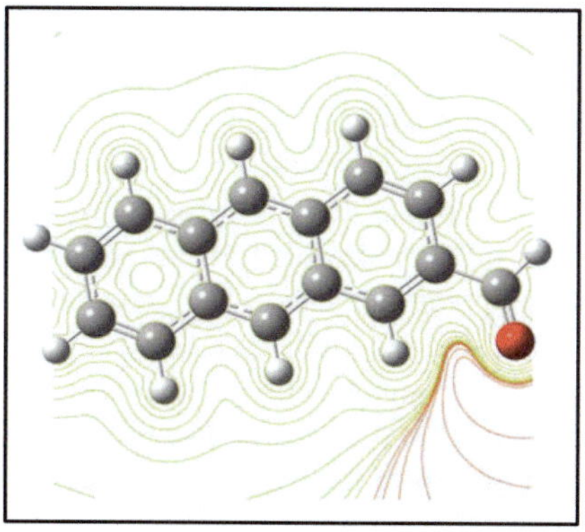

(b) Ar-CHO contour distribution

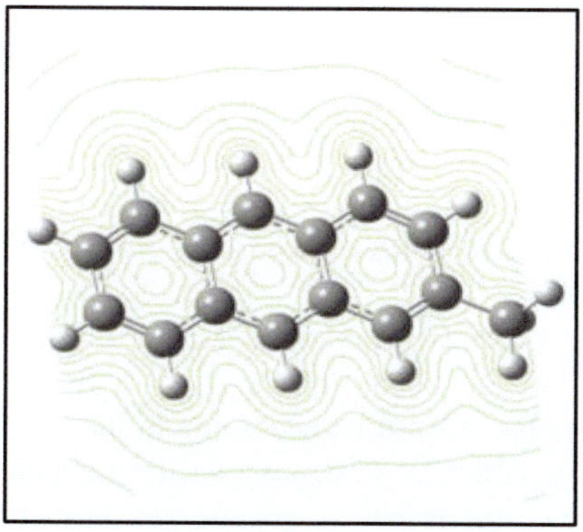

(c) Ar-CH3 contour distribution

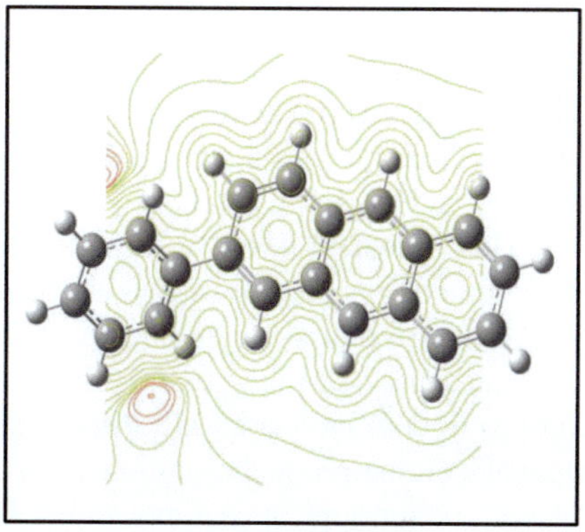

(d) Benzene ring side chain contour distribution

Fig. 3.4 Potential contour plots for the WYM molecular model

In order to change the proportion of different groups in coal molecules by solvent extraction, it is necessary to select an effective extractant according to the characteristics of the groups contained in coal molecules. Therefore, based upon the analysis results of the molecular structure characteristics of coal in Chap. 2, the extractants selected in this paper need to have different extraction capabilities for hydroxyl (–OH), carboxy groups, aliphatic side chains and aromatic hydrocarbons, respectively. Since hydroxyl (–OH) and carboxy groups are polar substances, while aliphatic hydrocarbons and aromatic hydrocarbons are non-polar substances (Lu et al. 2022), according to the principle of "similar dissolution" in the extraction principle, two polar extractants and two non-polar extractants should be selected to achieve the dispersion of these four groups. When choosing polar extractant, it is necessary to consider its non-covalent bonding ability with hydroxyl (–OH) and carbon oxygen groups. The stronger the bonding ability, the better the extraction effect. In addition to the principle of "similar dissolution", the structural differences of aliphatic hydrocarbons and aromatic hydrocarbons should be considered comprehensively when selecting polar extractants.

The molecular structure of ethylenediamine is shown in Fig. 3.5. It can be seen from the figure that the solvent molecule contains two amino groups, and the electrons in the molecule will be enriched in the amino group, so the solvent has a strong polarity. When it is in contact with coal molecules, the two amino groups in ethylenediamine molecule can form N–H–O hydrogen bonds with the –OH in coal molecules, making ethylenediamine have a stronger ability to form raextract on –OH than NMP, and can carry away a large number of coal molecular models containing –OH groups in the extraction process, thereby reducing the content of –OH in coal molecules.

The structure of N-methylpyrrolidone (NMP) is shown in Fig. 3.6. It can be seen from the figure that the pyrrole ring in NMP molecule contains a N atom with high electronegativity, so its molecular polarity is strong. Therefore, when NMP is in contact with coal molecules, the N atom in the intramolecular pyrrole ring can form a hydrogen bond with the O atom in the carbon oxygen group in the coal molecule (such as –CHO, and C–O–C), so as to realize the dissolution of the carbon oxygen group in the coal molecule by the extractant, so as to achieve the purpose of reducing the carbon oxygen group in the coal molecule.

Petroleum ether, as a common non-polar extractant, is composed of many different hydrocarbons. According to the principle of organic chemistry, the aliphatic hydrocarbons in coal molecules are usually non-polar. Therefore, according to the principle of "similar phase dissolution", petroleum ether can better dissolve the aliphatic hydrocarbon groups in coal molecules. Because the solubility of petroleum ether to non-polar substances increases with the increase of its relative molecular weight, the long-chain petroleum ether has better solubility to aliphatic hydrocarbon groups than the short-chain petroleum ether. The petroleum ether whose principal component is n-hexane is the petroleum ether extractant with the largest relative molecular weight.

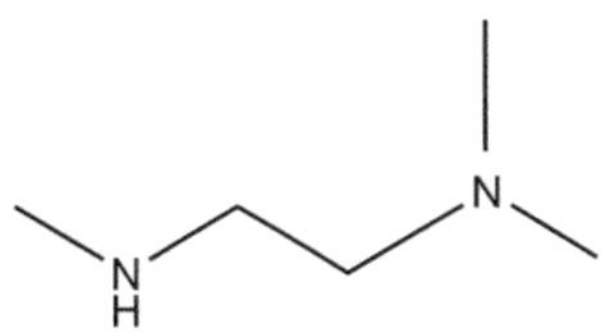

Fig. 3.5 Molecular structure of ethylenediamine

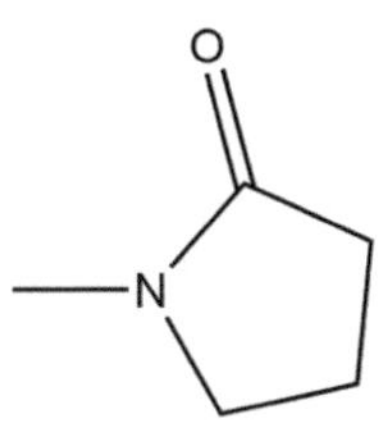

Fig. 3.6 Molecular structure of NMP

Fig. 3.7 Molecular structure of n-hexane

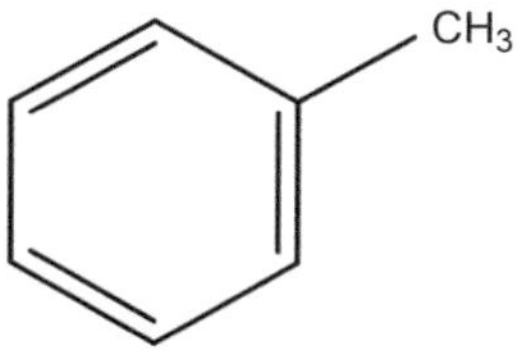

Fig. 3.8 Molecular structure of toluene

Therefore, n-hexane petroleum ether is selected as the extractant of discrete aliphatic hydrocarbon group in this paper, and its molecular structure is shown in Fig. 3.7.

Toluene, a non-polar substance, can extract hydrocarbon groups in coal molecules as effectively as petroleum ether. However, because the benzene ring structure in toluene molecule can be π–π stacked with the aromatic ring in the aromatic hydrocarbon group, a strong interaction force is formed. Therefore, compared with petroleum ether, toluene has a stronger ability to disperse aromatic hydrocarbon groups in coal molecules, which can effectively reduce the content of aromatic hydrocarbon groups in coal molecules. Therefore, toluene was selected as an organic solvent for extracting aromatic hydrocarbon groups in coal molecules, and its molecular structure was shown in Fig. 3.8.

3.2.2 *The Extraction Agent Electrostatic Potential Distribution Characteristics*

Upon previous studies and the basic principles of extraction, extractants with different extraction capabilities were selected for aromatic, –OH, extraction agents with different extraction capacity were selected for the oxygen-containing groups and aliphatic groups except –OH, in order to achieve the purpose of stripping side chain molecular models of different groups in coal, so as to study the reaction mechanism of different groups in the oxidation process.

Figure 3.9 shows the electrostatic potential diagram of the side chain divided into ethylenediamine ($C_2H_8N_2$), NMP (mainly C_5H_9NO), petroleum ether (mainly hexane) and toluene ($C_6H_5CH_3$).

As shown in Fig. 3.9, the pronounced potential distribution on the surface of different extractants differs greatly. Generally speaking, the electrostatic potential of n-hexane, the main component of petroleum ether (hereinafter referred to as petroleum ether by n-hexane, the main component of petroleum ether) is mainly distributed uniformly with positive electric potential. NMP (the main component

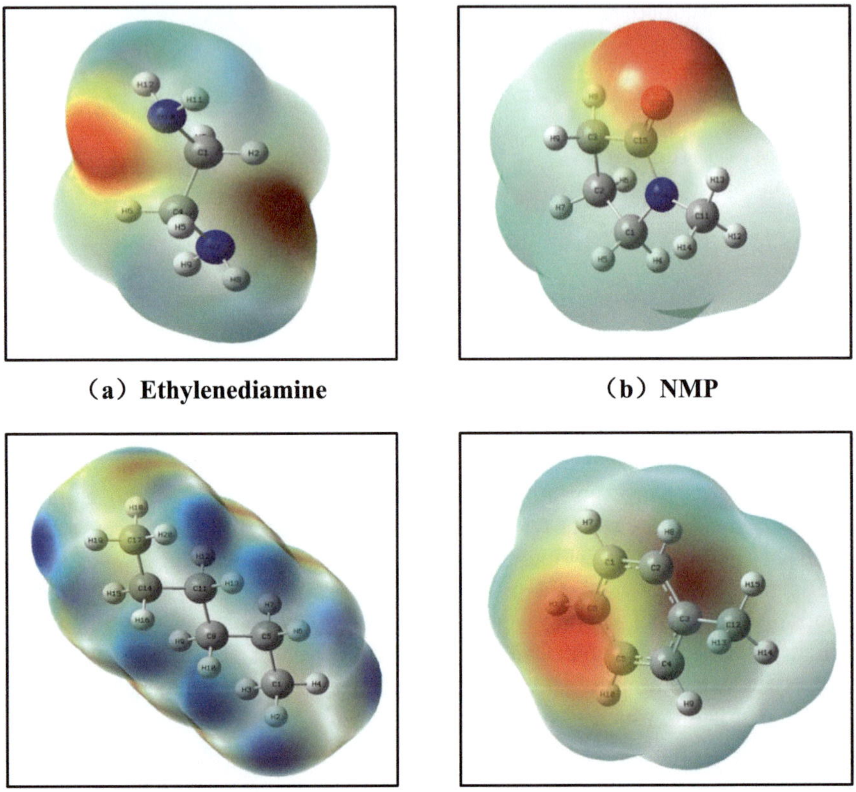

(a) Ethylenediamine **(b) NMP**

(c) N-hexane **(d) Toluene**

Fig. 3.9 Surface electrostatic potential diagram of extractant

of NMP in NMP mixed solution) and ethylenediamine respectively have obvious electron enrichment regions.

N-hexane is a typical long-chain alkane organic molecule with weak polarity, with electron-deficient region on H atom and electron-rich region between H atom, but the electrostatic potential in this region is low. Ethylenediamine has two electron enrichment regions, which is caused by the N atom on the two amino groups in ethylenediamine molecule. The N atom is an atom with strong electronegativity. When H atom forms a covalent bond, it seizes the electrons of H atom, making itself an electron enrichment region with large negative potential. The negative potential region of NMP molecule is clustered around the O atom at the C=O bond, and the electronegativity is weakened because the N atom on its ring forms a conjugate structure with the C atom. Therefore, although NMP has two atoms with strong electronegativity, the negative potential region is only concentrated around the O atom. The electron enrichment region of toluene is located in the benzene ring, because the benzene ring is a conjugated system in which π electrons are shared to form a series of stable conjugated bonds. These π electrons form a stable electron density cloud on the benzene ring, making the benzene ring an electron enriched region. Compared with the benzene ring, the methyl group has a weak ability to attract electrons and becomes an electron donor. Therefore, the potential distribution of toluene shows a weak positive potential in the methyl region, while the C atom in the aromatic ring shows a strong negative potential.

3.3 Weak Interactions Between Coal Auto-Ignition Reactive Groups and Extractant Molecules

Weak Interaction (Zhou 2022) is a non-covalent interaction other than ionic bonding, and its nature is dominated by electrostatic attraction with a small amount of dispersive attraction. The nature of the extraction effect of organic solutions on coal sample molecules is manifested in the destruction of intermolecular bridge bonds between organic solvent molecules and coal sample molecules by weak interactions (including hydrogen bonding, dihydrogen bonding, π-hydrogen bonding, π–π stacking, van der Waals attraction, and so on) (Zhang et al. 2021; Gao and Yang 2017). The analysis of weak interactions can largely help researchers to screen effective organic solvents (Zhang et al. 2023) and predict the extraction effect of organic solvents on each group.

Approximate density gradient (ADG) analysis (Chen et al. 2016; Jiang 2023), as an extension of the most commonly used and classical weak interaction analysis method, topological analysis (AIM), can visualise the weak interactions between molecules (Ye et al. 2022). Based on the electrostatic potential analysis of the coal sample molecular model and organic solvent molecules in Sections I and II, the weak interactions between organic solvent molecules and coal sample molecules were calculated and analysed using ADG analysis (Liu 2022). YM molecules with different side chains were analysed for weak interactions with each of the four extractants (Li et al. 2022), and the optimised organic solvent molecules and coal sample models were added to a molecule group for the optimised frequency calculations using Gaussian 16 (Mao et al. 2021), after which the sign(λ_2)ρ(a.u.) of the weak interactions between the molecules were computed and plotted by Multiwfn and VMD software ADG scatter plots and filled isosurface maps. The Gaussian 16 calculation parameters were set as follows: Job Type: Opt + Freq; Method: Ground State, DFT, Default Spin, B3LYP; Basis set:6-311G(d). Multiwfn calculations were performed using ADG calculations (Wang et al. 2021).

3.3.1 *Weak Interaction of YM Reactive Groups with Extractants*

(a) **YM Reactive Group and Ethylenediamine Extractant**

Figure 3.10 shows the scatter plot and the colour-filled isosurface diagram of the molecular model of YM reactive group with –OH side chain and ethylenediamine extractant, from which it can be seen that there exists a dense spike located at about 0.022 in the region where sign(λ_2)ρ(a.u.) is greater than 0, which combined with the colour-filled isosurface diagram can be seen to be the effect of the site-blocking effect in the bicycle molecule; the dense spike located at about ±0.03 and about ±0.01 indicates that the ethylenediamine molecule has a close relationship with YM with the side chain of –OH group. The dense spikes at about ±0.03 and about ±0.01 indicate that there are substantial van der Waals forces between the ethylenediamine molecule and the YM reactive group molecular model with –OH groups on the side chains, corresponding to the weak interactions between the ethylenediamine-carrying H atoms and the H atoms on the YM reactive group shown in the filled isosurface plots.

There is an obvious spike peak at about −0.025 of sign(λ_2)ρ(a.u.) in the scatter plot, indicating that there is a strong attraction between the ethylenediamine molecule and the YM molecule with –OH side chain, and the analysis of the colour-filled isosurface diagram shows that the spike corresponds to the attraction formed by the H atom connected with the N atom and the O atom on the molecular model of the –OH side chain of the YM active group in the isosurface diagram. Since the value of the spike tip is much larger than −0.005, it is not considered that the attraction belongs to the category of van der Waals force, and the composition of O–H–N is in line with the conditions for the formation of hydrogen bonding, therefore, it is considered that ethylenediamine and –OH on the molecular model of the YM reactive group form O–H–N hydrogen bonding, which can be judged that ethylenediamine has a strong

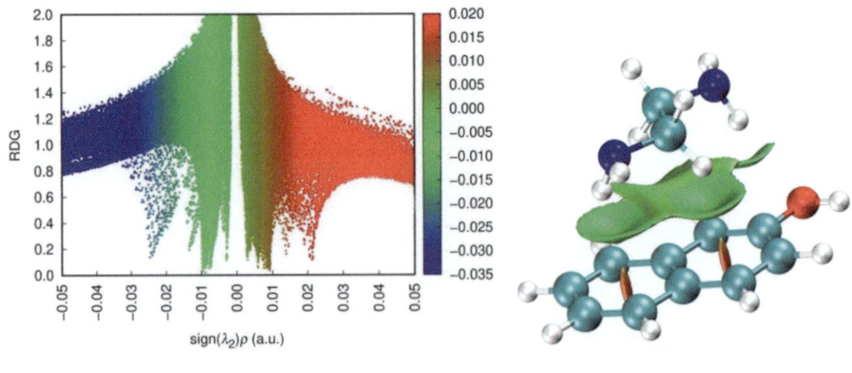

Fig. 3.10 Scatter plot and filled isosurfaces of YM molecular model with ethylenediamine extractant with –OH group in the side chain

bonding effect on –OH on the molecular model of the YM reactive group. OH on the model of YM active group molecule.

Figure 3.11 shows the scatter plots and filled isosurface plots of YM reactive group molecular model with –CHO side chain and ethylenediamine extractant. From the plots, it can be seen that all YM reactive group molecular models are removed and there exists a spike with sign(λ_2)ρ(a.u.) value of about 0.022, and there appears a new spike with sign(λ_2)ρ(a.u.) value of about 0.012, which can be seen from the isosurface plot analysis. Combined with the analysis of the isosurface plots, it can be seen that this spike corresponds to the site-blocking effect of the O on the carbon–oxygen group of the YM reactive group molecular model bonded to the neighbouring H atom of the YM reactive group molecular model, and the existence of this site-blocking effect indicates that –CHO and the YM reactive group molecular model can be easily fractured from the connected C–C; and removing the sign(λ_2)ρ(a.u.) value of about −0.01 spike, no weak interactions are evident on the left side of the scatter plot. According to the colour-filled contour plot, it can be seen that the O atom on the –CHO side chain and the H atom attached to the N atom of ethylenediamine conform to the formation of a hydrogen bond constitutively, but due to the limitation of the weak interactions between the molecular models of ethylenediamine and owing the YM reactive moiety, although this H atom forms a hydrogen bond with the O atom, O–H–N, it fails to be manifested in the y = 0.5 contour plot due to the relatively long relative distance. Taken together, the ethylenediamine extractant has only a weak attraction to the –CHO side chain on the molecular model of the YM reactive group, and thus the extraction ability of the ethylenediamine extractant for –CHO is not as good as that for –OH.

Figure 3.12 shows the scatter plot and colour-filled isosurface plots of the molecular model of YM reactive moiety with ethylenediamine extractant with $-CH_2-$ side chain, from which it can be seen that there are only three obvious spikes in the scatter plot of the molecular model of YM reactive moiety with ethylenediamine extractant with sign(λ_2)ρ(a.u.) values of about ±0.05 and 0.022, respectively, and

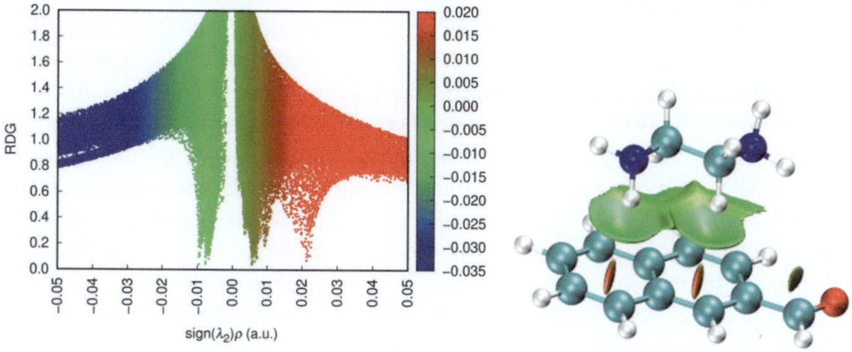

Fig. 3.11 Scatter plot and filled isosurfaces of YM molecular model with ethylenediamine extractant with –CHO group in the side chain

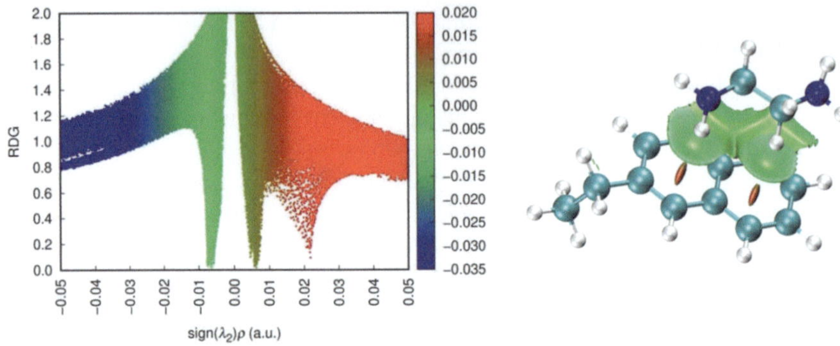

Fig. 3.12 Scatter plot and filled isosurfaces of YM molecular model with –CH_2– group in side chain with ethylenediamine extractant

these three spike correspond to the van der Waals forces between the YM reactive group molecular model and ethylenediamine and the site-barrier effect in the YM reactive group on the isosurface diagram, respectively, and it is clearly seen that the ethylenediamine molecules do not have any interactions with the –CH_2– side chain on the YM reactive group molecular model, which suggests that the ethylenediamine extractant does not have any extractive effect on the –CH_2– side chain on the YM reactive group molecular model This conclusion is also in line with the principle that extractants will choose to dissolve substances of the same polarity as them (Li et al. 2019).

Figure 3.13 shows the scatter plot and filled isosurface plots of the YM reactive group molecular model with aromatic side chains against the ethylenediamine extractant, from which it can be seen that compared with the ADG plot of the YM molecular model with –CH_2– side chains against ethylenediamine, two new spikes appear outside of the spike with the sign(λ_2)ρ(a.u.) value of this scatter plot of approximately ±0.005, which are spike are the same at ±0.01, but with different peak heights, which suggests that there are two different weak interactions between the molecular model of the YM reactive group with aromatic groups in the side chain and the ethylenediamine molecule or within each of them, which can be analysed in conjunction with the isosurface plots to conclude that these two weak interactions occur in the van der Waals forces generated between ethylenediamine and YM on the aromatic groups and YM molecules, respectively, and that The H–π bond is formed between the N atom on the ethylenediamine. In addition, the C atom on the aromatic ring, but since sign(λ_2)ρ(a.u.) is close to the range of the van der Waals force, the weak interaction force corresponding to this spike produces a weak attraction, and it is difficult for the side chain to fracture from the aromatic side chain to the YM molecule due to dispersive forces on both sides of the bridge bond between the two molecules. Upon the above analysis, the ethylenediamine extractant has basically no extraction ability for the aromatic side chain on the molecular model of the YM reactive group.

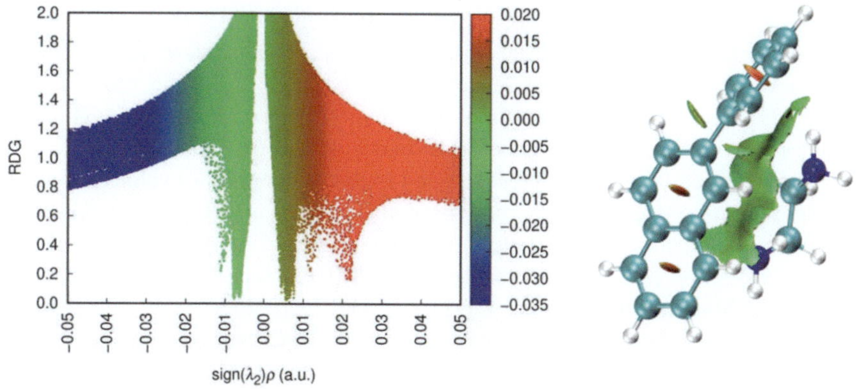

Fig. 3.13 Scatter plot and filled isosurfaces of YM molecular model with aromatic groups in the side chain versus ethylenediamine extractant

(b) **YM Reactive Group with NMP Extractant**

Figure 3.14 shows the scatter plot and the filled isosurface map of the YM molecule group with –OH side chain and NMP extractant, from which it can be seen that there are two spikes in the red region, and the values of the tip sign(λ_2)ρ(a.u.) are about 0.022 and 0.042, respectively, which indicate that there exist two positions between the YM molecule carrying oxygen-containing groups and NMP or within the two molecules that have strong and strong repulsion effect, combined with the analysis of the colour-filled isosurface plot with y = 0.5, it is known that this is due to the site-blocking effect in the piperidine in the NMP molecule and the biphenyl ring of the YM molecule; the presence of spike at the sign(λ_2)ρ(a.u.) value of about –0.08 indicates that the YM molecule carrying the –OH moiety has a strong repulsive effect on the NMP molecule and NMP molecule.

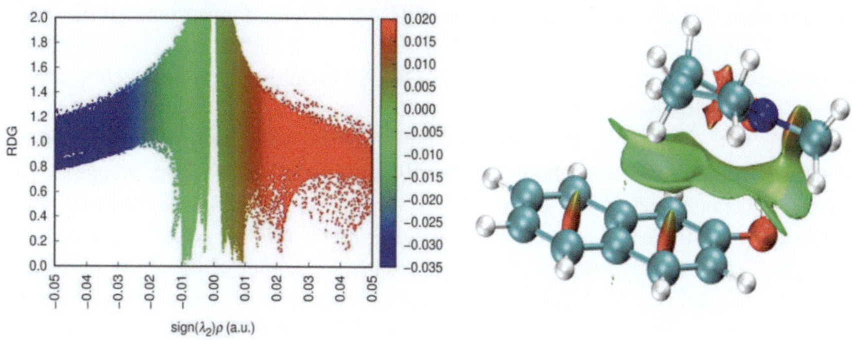

Fig. 3.14 Scatter plot and filled isosurface of YM molecular model carrying –OH groups with NMP extractant

–OH group between YM molecule and NMP, combined with the analysis of the filled isosurface plot with isosurface y = 0.5, it can be seen that, in addition to the van der Waal's force between the H atoms on the benzene ring of the YM molecule and the H atoms of NMP, there is a N–H–O hydrogen bond between the H atoms on the methyl group carried by the NMP molecule and the O atoms on the –OH side chain carried by the YM molecule; and in the sign(λ_2)ρ(a.u.) values are much smaller region no longer exists a more pronounced spike, indicating that almost no hydrogen bond can be formed between the bis-benzene ring of the YM molecule and the NMP molecule. This suggests that the NMP extractant has some extraction ability for YM molecules carrying –OH groups, and the H atoms on its methyl group and the O atoms on the –OH side chain can form N–H–O hydrogen bonds.

Figure 3.15 shows the scatter plot and filled isosurfaces of the YM molecule moiety with carbon–oxygen group in the side chain and NMP extractant, in the scatter plot, it can be seen that there are two spikes in the red region with the tip sign(λ_2)ρ(a.u.) values of about 0.022 and 0.042, which can be seen from the analysis of the previous subsection, which is caused by the piperidinium in the NMP molecule and in the biphenyl ring of the YM molecule respectively The presence of the site-blocking effect; and a distinct spike at sign(λ_2)ρ(a.u.) of −0.015, which is more pronounced and has a smaller sign(λ_2)ρ(a.u.) value compared with the spike at sign(λ_2)ρ(a.u.) −0.08 in the scatter plot of the YM molecule carrying –OH versus the NMP, suggests that the YM molecule carrying the This peak is more pronounced and the sign(λ_2)ρ(a.u.) value is smaller, suggesting that there is a van der Waals attraction between the YM molecule carrying the carbon–oxygen group and the NMP molecule, which is stronger than that of the NMP molecule for –OH. Combined with the analysis of the filled isosurface plot with y = 0.5, it can be seen that the C atom on the –CHO side chain of the YM molecule forms a van der Waals attraction with an H atom of the methyl group in the NMP molecule; and there no longer exists a spike on the left side of the sign(λ_2)ρ(a.u.) value of −0.015, so it can be clarified that the extraction capacity of the NMP on the YM molecule is generated by the van der Waals attraction formed only between the methyl group and –CHO in the NMP molecule, which is stronger than that of NMP molecules on –OH. CHO, which indicates that the NMP extractant has extraction ability for YM molecules with carbon–oxygen side chains and is stronger than its extraction ability for the YM reactive group molecular model with –OH side chains.

Figure 3.16 show the scatter plots and colour-filled isosurfaces of YM molecular moieties with aliphatic side chains and NMP extractants, from which it can be seen that the spikes characterizing the site-blocking effect in the rings of YM and NMP molecules are still present in the red region, with sign(λ_2)ρ(a.u.) values of about 0.022 and 0.042, respectively; and in comparison with the scatter plots of the first three different side chains of YM molecules with NMP, there is no distinct difference in the green area except for the sign(λ_2)ρ(a.u.) value of about 0.022 and 0.042. Compared with the scatter plots of the first three different side chains of YM molecules with NMP, there are no more spikes in the green except the spike with the sign(λ_2)ρ(a.u.) value of about −0.008, and combined with the analysis of the colour-filled isosurface plots of y = 0.5, it can be seen that there are almost no weak interaction forces between

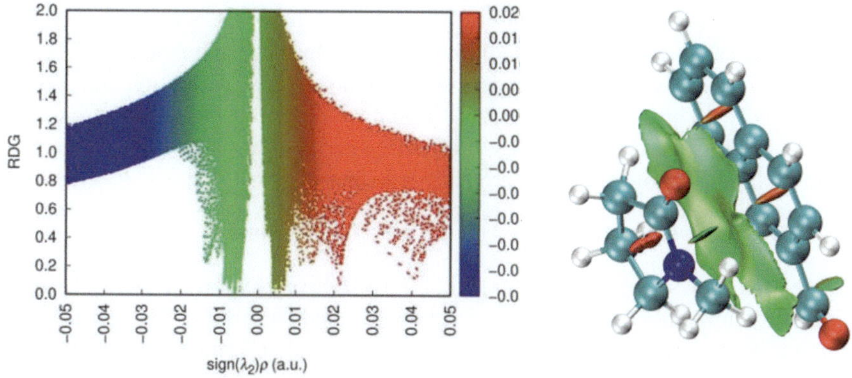

Fig. 3.15 Scatter plot and filled isosurfaces of YM molecular model of NMP extractant with carbon–oxygen groups in the side chain

the aliphatic group side chains and NMP molecules as well, and it can be seen that the NMP extractant has almost no extracting effect on the YM molecules with the aliphatic group side chain as well. There is no extraction effect of NMP extractant on YM molecules with aliphatic side chains.

Figure 3.17 shows the scatter plot and colour-filled isosurfaces of the YM molecular moiety with aromatic side chains and NMP extractant, from which it can be seen that the red region still has a spike with sign(λ_2)ρ(a.u.) values of about 0.022 and 0.042, respectively, which is due to the site-blocking effect in the rings of the YM molecule and the NMP molecule; observing the green region of the scatter plot, it can be seen that A new spike appears at sign(λ_2)ρ(a.u.) of about 0.014, but it is small compared with the spikes at the YM molecule carrying –OH and carbon–oxygen groups. Combined with the analysis of the filled isosurface plot at $y = 0.5$, it can be seen that the van der Waals force corresponding to the newly appeared spike occurs at the bridge bond between the benzene ring side chain and the biphenyl ring,

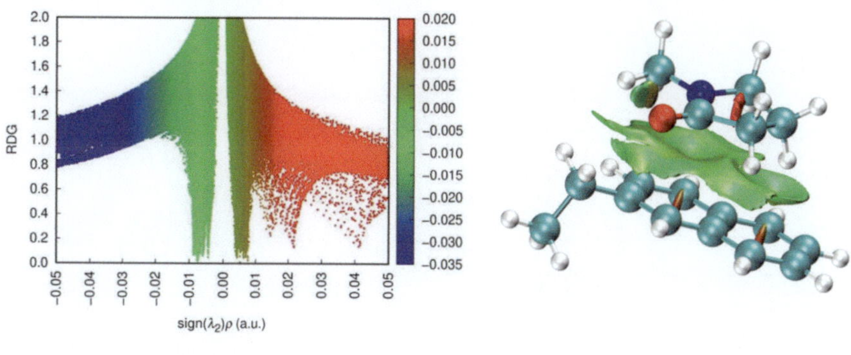

Fig. 3.16 Scatter plot and filled isosurfaces of YM molecular model with aliphatic groups in the side chain versus NMP extractant

whereas no new weak interactions appear at the other positions, which suggests that the aromatic side chain will be more difficult to be disconnected than the –OH side chain, the carbon–oxygen group side chain, and the aliphatic side chain. Since there is essentially no attraction between the aromatic side chains of the YM molecules and the NMP extractant molecules, the NMP extractant has little or no extraction effect on the YM molecules with aromatic side chains.

(c) YM Reactive Group and Petroleum Ether Extractant

Figure 3.18 shows the molecular model of YM reactive group with –OH side chain with petroleum ether extractant scatter plot and filled isosurface. The spike of this scatter plot is divided into three parts, and the sign(λ_2)ρ(a.u.) values are about ±0.09 and −0.22, respectively. Based on the spike distribution in the scatter plot alone, it is believed that the petroleum ether does not have any extraction effect on the YM reactive group. However, combining the isosurface plots, it can be seen that there is a van der Waals force between the H atoms on the terminal methyl group of ethylenediamine and the O atoms of the –OH side chain, but due to the weak interaction force, it overlaps with the spike with a sign(λ_2)ρ(a.u.) value of about −0.09 in the scatter plot, and thus the spike corresponding to the van der Waals force is not found in the scatter plot directly. Due to the low attraction strength of this van der Waals force, it is believed that the petroleum ether does not have any extracting effect on the YM active group. strength of the attraction is low, the petroleum ether extractant is considered to be less capable of extracting YM reactive groups with –OH side chains.

Figure 3.19 shows the molecular model of YM reactive moiety with –CHO side chain with petroleum ether extractant scatter plot and filled colour isosurface plot, by observing the scatter plot, it can be found that there exists only one spike with the sign(λ_2)ρ(a.u.) value of about 0.012, except the three spikes with sign(λ_2)ρ(a.u.) values of about ±0.09 and −0.22, respectively. And this spike corresponds to the site-blocking effect between –CHO and the YM reactive group on the filled isosurface map. The isosurface plots of petroleum ether with –CHO side chain for the molecular

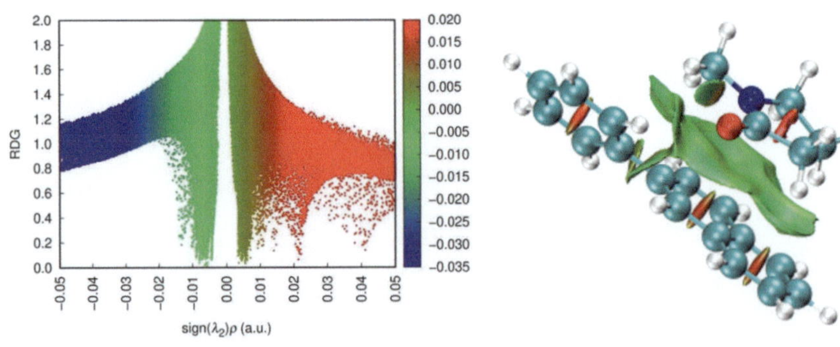

Fig. 3.17 Scatter plot and filled isosurfaces of YM molecular model with aromatic side chains versus NMP extractant

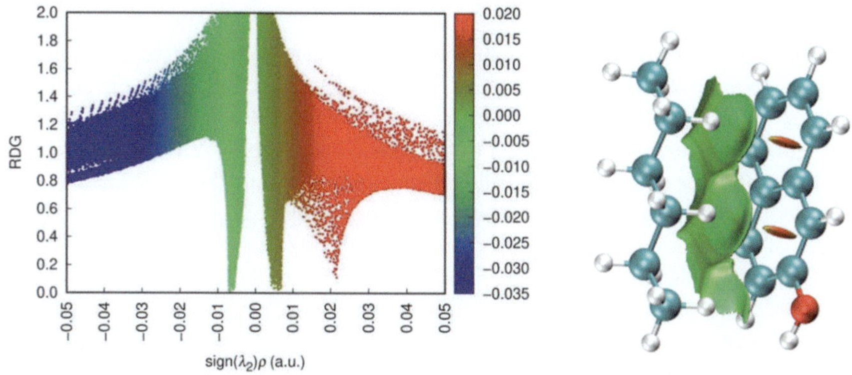

Fig. 3.18 Scatter plot and filled isosurfaces of YM molecular model with –OH side chain versus petroleum ether extractant

model of YM reactive group are almost the same as those of ethylenediamine with –CHO side chain for the molecular model of YM reactive group, which indicates that petroleum ether lacks the ability to extract YM reactive group with –CHO side chain, just as the ethylenediamine extractant does.

Figure 3.20 shows the molecular model of YM reactive group with $-CH_2-$ side chain with petroleum ether extractant scatter plot and filled isosurface plots. Observation of the scatter plots reveals that there exists a prime dense spike at the sign(λ_2)ρ(a.u.) value of foughyt 0.08, which suggests that there exists a weak interaction between the petroleum ether molecule and the YM reactive group molecule with the $-CH_2-$ side chain. Combined with the isosurface diagram, it can be seen that there is a dispersion force between the terminal H atoms of the petroleum ether molecule and the H atoms on the $-CH_2-$ side chain of the YM molecule, and since this spike is larger than the spike to the right of it with a sign(λ_2)ρ(a.u.) value of approximately −0.07, it can be judged that this interaction is stronger than the H–π bond between the

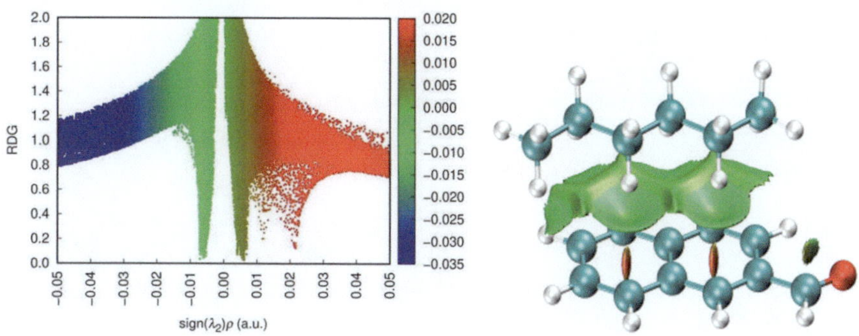

Fig. 3.19 Scatter plot and filled isosurface of YM molecular model with petroleum ether extractant with –CHO side chain

petroleum ether molecule and the aromatic ring of the YM molecule, which suggests that there is a weak interaction of the petroleum ether on the YM reactive moiety molecule with a –CH_2– side chain of –CH_2–. This indicates that petroleum ether has a strong extraction ability for YM reactive groups with –CH_2– chain.

Figure 3.21 shows the scatterplot of the molecular model of YM reactive moiety with aromatic side chain versus petroleum ether extractant and the color-filled isosurface plots. As can be seen from the figure, compared to the scatterplot of the molecular model of YM reactive moiety with petroleum ether extractant with the side chain of –CH_2–, there is a newly appeared spike in this scatterplot at the sign(λ_2)ρ (a.u.) value of about −0.01 and is smaller compared to the spike located at the sign(λ_2)ρ (a.u.) value of about −0.01 is smaller. Combined with the analysis of the color-filled isosurface plots, it can be seen that the spike present at the sign(λ_2)ρ(a.u.) value of about −0.01 corresponds to the site-blocking effect in the side chain of the aryl ring, while the spike at the sign(λ_2)ρ (a.u.) value of about −0.01 corresponds to the formation of the H–π bond between the methyl group at the end of the petroleum ether molecule and the aryl ring, and since this van der Waals force corresponds to the spike on the scatterplot Since the van der Waals force on the scattering diagram corresponds to a spike smaller than the dispersion force spike formed between the petroleum ether and the YM reactive group –CH_2–, the petroleum ether has a certain extraction effect on the YM molecule for the YM reactive group with an aromatic side chain, which is stronger than that for the YM reactive group with a side chain of –OH and carbon–oxygen groups.

(d) YM Reactive Group with Toluene Extractant ADG Analysis

Figure 3.22 shows the molecular model of the YM reactive group with –OH side chain with toluene extractant scatter plot and filled isosurface, by observing the scatter plot, it can be seen that by removing the spike with sign(λ_2)ρ(a.u.) value of about ±0.005 corresponding to the van der Waals force between the two molecules, there exists also a larger spike with sign(λ_2)ρ(a.u.) value of about −0.07 which indicates the existence of strong van der Waals forces between the two molecules. Combined

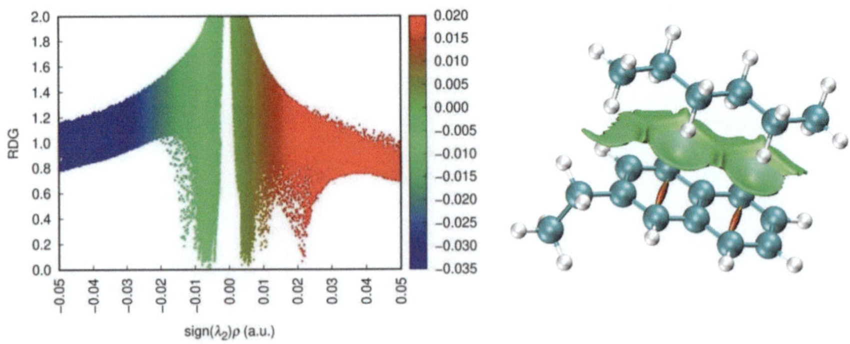

Fig. 3.20 Scatter plot and filled isosurfaces of YM molecular model with –CH_2– side chain versus petroleum ether extractant

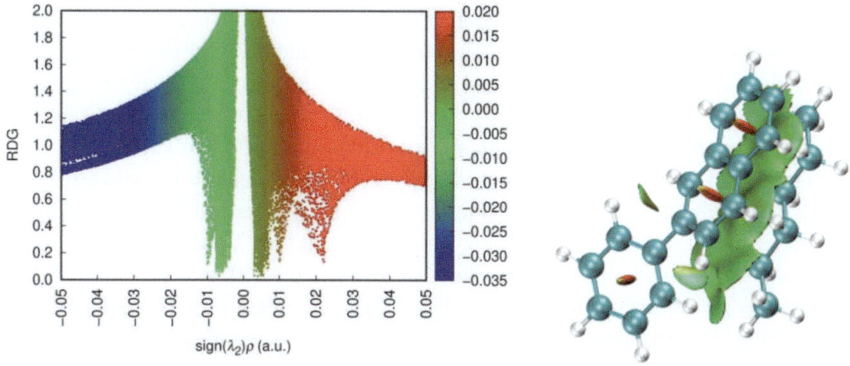

Fig. 3.21 Scatter plot and filled isosurfaces of YM molecular model with aromatic side chains versus petroleum ether extractant

with the analysis of the isosurface plots, it can be seen that the existence of van der Waals forces between the two molecules is due to the formation of π–π stacking on toluene with the aromatic rings of the YM molecules, and there are no other weak interactions between the side chains of the YM reactive group molecules, –OH and toluene, so that the toluene extractant has a poor extraction effect on the YM reactive group with the side chain of –CHO.

Figure 3.23 shows the molecular model of the YM reactive group with a side chain of –CHO with the toluene extractant scatter plot and filled isosurfaces, and analysis of the scatter plot shows that, removing the spike at the sign(λ_2)ρ(a.u.) value of rougly −0.007 corresponding to a weak interaction force between the two molecules, there also exists a sign(λ_2)ρ(a.u.) value of about 0.012 at the spike, which indicates the existence of a weaker interaction force between or within the two molecules. By observing the isosurface diagrams, it can be found that the hydrogen

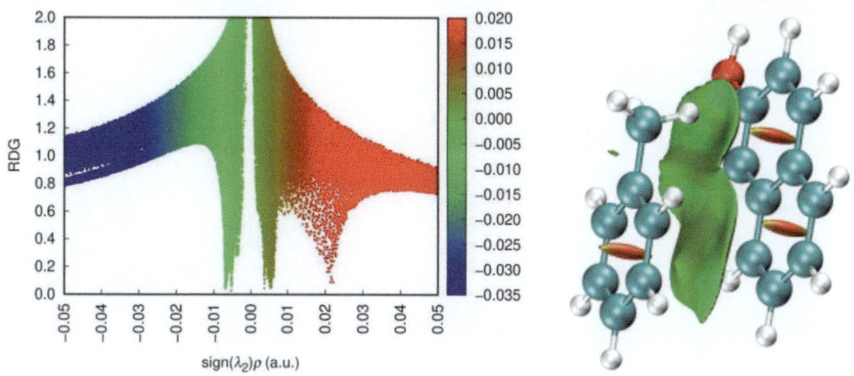

Fig. 3.22 Scatter plot and filled isosurfaces of YM molecular model with –OH side chain versus toluene extractant

bond corresponding to the spike is located between the O atom at the end of the side chain and the neighbouring H atom, and the –CHO side chain of the YM molecule only has a weak van der Waals force with the methyl group of the toluene molecule, and there is no other interaction force, which indicates that toluene has a general effect on the YM reactive group with the side chain of –CHO extracted.

Figure 3.24 Scatterplot and filled isosurface of the molecular model of the YM reactive group with $–CH_2–$ side chain versus the toluene extractant. By looking at the scatterplot, it can be seen that there is only one spike with a sign(λ_2)ρ(a.u.) value of about -0.008 in the direction less than sign(λ_2)ρ(a.u.) $= 0$, which corresponds to the toluene-YM molecule-to-molecule interaction, and there is no mutual attraction beyond that. Observation of the filled isosurface plots similarly shows that all the interactions of toluene with the YM reactive group with $–CH_2–$ side chain are present between the YM aromatic ring and the toluene aromatic ring, indicating that the toluene extractant does not interact weakly with $–CH_2–$ on the YM molecule, and thus the toluene extractant is weakly capable of extracting the YM molecule with an aromatic side chain.

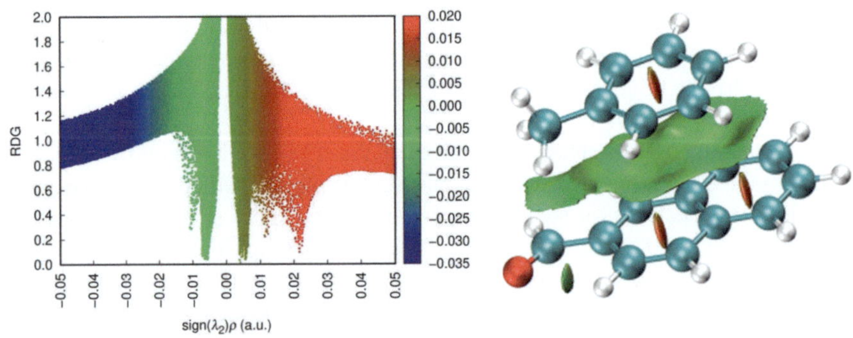

Fig. 3.23 Scatter plot and filled isosurfaces of YM molecular model with –CHO side chain versus toluene extractant

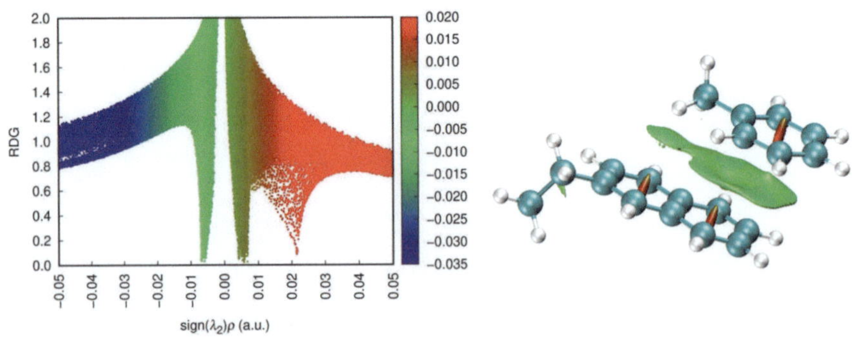

Fig. 3.24 Scatter plot and filled isosurfaces of the YM molecular model with $–CH_2–$ side chain versus toluene extractant

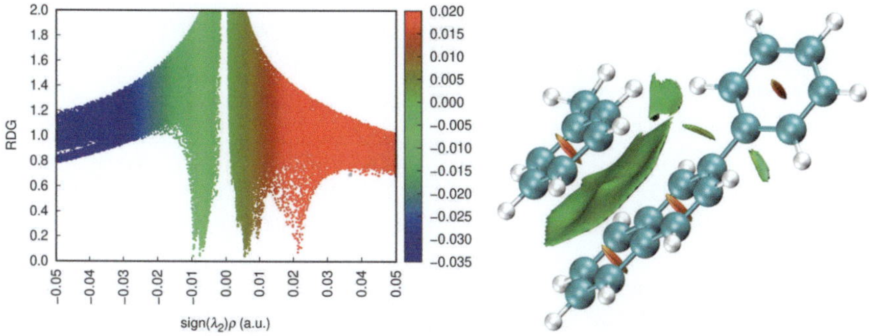

Fig. 3.25 Scatter plot and filled isosurfaces of YM molecular model with aromatic side chains versus toluene extractant

Figure 3.25 shows the scatter plots and filled isosurfaces of the YM reactive group molecular model with aromatic side chains versus the toluene extractant, and it can be noticed just by looking at the scatter plots that the ADG scatter plots of toluene with the YM reactive group molecular model with aromatic side chains are quite different from those of toluene with the YM reactive group molecular model with the side chains of the other moieties. Excluding the spike with $sign(\lambda_2)\rho(a.u.) > 0$ corresponding to the site-barrier effect in the aryl ring, the spike at $sign(\lambda_2)\rho(a.u.) < 0$ is significantly more electron-dense and appears to have a spike with a $sign(\lambda_2)\rho(a.u.)$ value of about −0.01, suggesting that there exists a significant van der Waals force between the two molecules. Combined with the isosurface diagram, it can be seen that this van der Waals force is located between the aromatic side chain of the YM molecule and the toluene aromatic ring. Therefore, the toluene extractant has a better extraction effect on YM molecules with aromatic side chains.

The mode of interaction between four different side chain YM reactive groups and four different extractant molecules is shown in Table 3.2.

As can be seen from the table, the weak interaction forces of the extractants on the coal molecules are mainly manifested as different strengths of hydrogen bonding, and van der Waals forces, in which ethylenediamine has a stronger extraction effect for –OH, NMP has a better extraction effect for –CHO, toluene extraction is more targeted at the aromatic group, and petroleum ether has a better extraction effect on the aliphatic group. By comparing the interaction forces and spike tip positions of coal molecules carrying different side chains with different extractants, it can be found that the ethylenediamine molecule can form O–H–N hydrogen bonding with coal molecules with –OH side chains, while the weak interaction forces with coal molecules with other side chains are smaller than those with –OH molecules, both in terms of the number and the size of the spikes on the scatter plot. NMP can form van der Waals force with the coal molecule model with –CHO side chain, and the spikes corresponding to the mutual attraction are more and denser than the spikes corresponding to –OH side chain YM molecule. NMP can form van der Waals forces with the coal molecular model with –CHO side chains, which corresponds to

Table 3.2 Interaction modes between different side chain YM active groups and different extractant molecules

Reactive radical	Spike tip position			
	Ethylenediamine	NMP	Petroleum ether	Methylbenzene
–OH	−0.003; −0.01; −0.025	−0.002; −0.008;, −0.012	−0.005	−0.005; −0.007
	0.005; −0.02	0.005; 0.009; 0.014	0.005; 0.022	0.005
–CHO	−0.008; −0.01	−0.005; −0.015	−0.005	−0.001; −0.007;
	0.005; 0.01	0.005; 0.022	0.005; 0.012	0.005; 0.012
$–CH_2–$	−0.007	−0.005	−0.06; −0.012	−0.005
	0.005	0.005; 0.015	0.005; 0.01	0.005
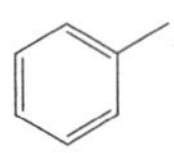	−0.005; −0.011	−0.006; −0.013	−0.006; −0.01	−0.008; −0.01
	0.005; 0.012	0.005; 0.01	0.005; 0.01	0.005; 0.009

more inter-attracting spikes and is more dense. Petroleum ether has dispersion force with toluene and aliphatic side chains of YM reactive group, but the H–π bond with aliphatic group has a larger spike, and its extraction ability for different side chain YM molecules is ranked as aliphatic side chain YM molecules > aromatic side chain YM molecules > –OH side chain YM molecules > carbonylic acid group side chain YM molecules. Toluene molecules can form van der Waals forces with YM molecules whose side chains are toluene, while there is no mutual attraction to the side chains of other YM reactive groups, and its extraction ability for different side chain YM molecules is ranked as aromatic side chain YM molecules > carbon–oxygen group side chain YM molecules > aliphatic side chain YM molecules > –OH side chain YM molecules.

3.3.2 *Weak Interaction of WYM Active Group with Extractant*

(a) **WYM Reactive Group with Ethylenediamine Extractant**

Figure 3.26 shows the scatter plot and color-filled isosurface plots of the molecular model of the WYM reactive group with –OH side chain and ethylenediamine extractant, from which it can be seen that there exists a small and low-density spike at the sign(λ_2)ρ(a.u.) value of −0.012. Combined with the analysis of the isosurface plots, it can be seen that the spike corresponds to the N–H–O hydrogen bonding between the molecules of ethylenediamine and WYM reactive group, but because the H atom is far away from the oxygen atom, this hydrogen bonding corresponds to low mutual attraction. The N–H–O hydrogen bond between ethylenediamine and WYM reactive

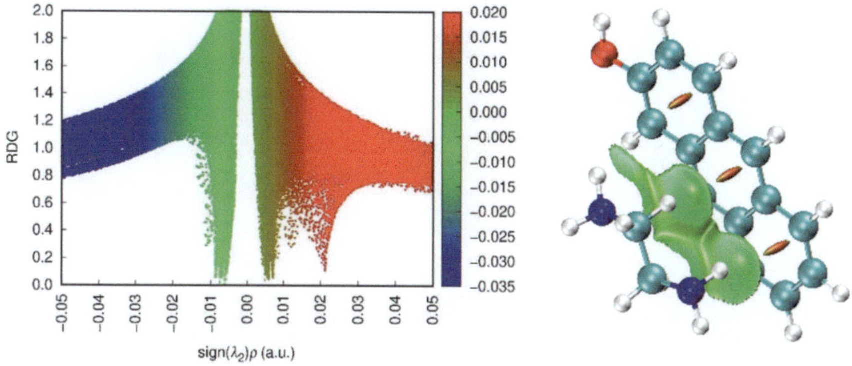

Fig. 3.26 Scatter plot and color-filled isosurfaces of WYM molecular model with ethylenediamine extractant with –OH energy group in the side chain

group molecules, but the H atom is far away from the oxygen atom, so the hydrogen bond corresponds to a low mutual attraction. However, compared with the ADG plots of ethylenediamine with WYM molecules containing side chains of other groups, ethylenediamine has a certain extraction effect on WYM reactive groups with –OH side chains.

Figure 3.27 shows the molecular model of the WYM reactive group with a carbon–oxygen group on the side chain with ethylenediamine extractant scatterplot and color-filled isosurface plots. It can be found by observing the scatterplot that, in addition to the spike corresponding to van der Waals interactions at the sign(λ_2)ρ(a.u.) value of about ±0.005, there is also a spike at the sign(λ_2)ρ(a.u.) value of about −0.06 to 0.08. This spike corresponds to the formation of H-π bonds between O atoms on the carbon–oxygen group and H atoms on the neighboring C atoms in the isosurface plot and is not a weak mutual attraction between the two molecules. There also exists a spike which corresponds to the H–π bond formed between the O atom on the carbon–oxygen group and the H atom on the neighboring C atom in the isosurface diagram, which is not a weak mutual attraction between the two molecules. Besides this spike, there are no other spikes with smaller values of sign(λ_2)ρ(a.u.), so ethylenediamine has poor extraction ability for WYM reactive groups with carbon–oxygen groups in the side chain.

Figure 3.28 shows the scatter plots and color-filled isosurfaces of the WYM reactive group molecular model with ethylenediamine extractant with aliphatic side chains. By comparing the ADG scatter plots of ethylenediamine with the WYM reactive group molecular model with aliphatic side chains, it can be seen that similar to the ADG scatter plots of the ethylenediamine with the WYM reactive group molecular model with aromatic side chains, the sign(λ_2)ρ (a.u.) values are Negative spikes are mainly concentrated between −0.005 and −0.008, and a sign(λ_2)ρ(a.u.) value of about −0.08 suggests that there may be a weak interaction force between the ethylenediamine molecule and the aliphatic group. Combined with the isosurface plots, it can be seen that this spike corresponds to the dispersion force formed

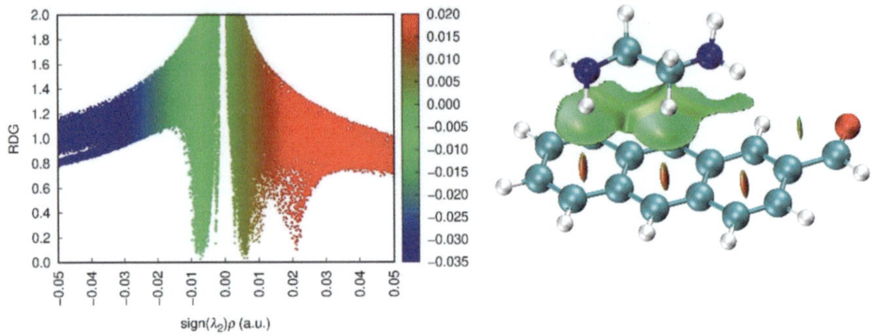

Fig. 3.27 Scatter plot and filled isosurfaces of WYM molecular model with ethylenediamine extractant with carbon–oxygen groups in side chains

between the aliphatic side chain and ethylenediamine, which has a stronger interaction force than the H–π bond formed between ethylenediamine and the aromatic side chain. However, apart from this dispersion force, there are no other spikes with smaller sign(λ_2)ρ(a.u.) values, suggesting that the spike with a sign(λ_2)ρ(a.u.) value of about −0.008 is the only weak mutual attraction between the two molecules apart from van der Waals interactions. Therefore, ethylenediamine has an average extraction ability for WYM reactive groups with aliphatic side chains, but is stronger than for WYM reactive groups with aromatic side chains.

Figure 3.29 shows the molecular model of the WYM reactive moiety with aromatic side chain as an aromatic moiety with the scatter plot and color-filled isosurface plots of the ethylenediamine extractant. By observing the scatter plots, it can be seen that the spikes with negative sign(λ_2)ρ(a.u.) values are mainly concentrated in the range of −0.005 to −0.01, and the spike with the sign(λ_2)ρ(a.u.) value of about −0.01 suggests that the Ethylenediamine molecule and the side chain is an aromatic group there may be a weak interaction force, combined with the isosurface plot it can be seen that the spike corresponds to the hydrogen bond formed between the H atom

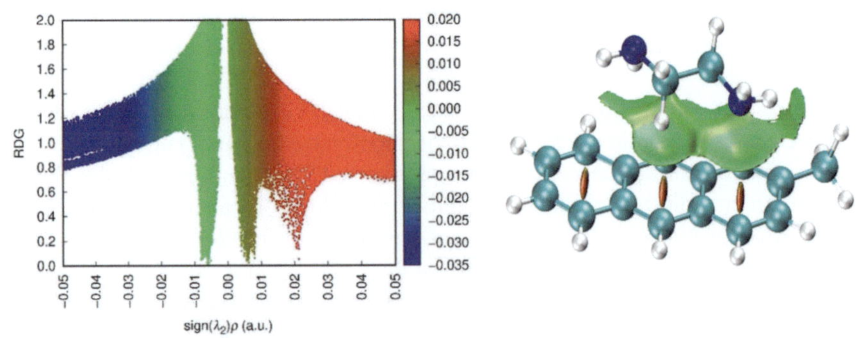

Fig. 3.28 Scatter plot and filled isosurfaces of WYM molecular model with ethylenediamine extractant for aliphatic groups in the side chain

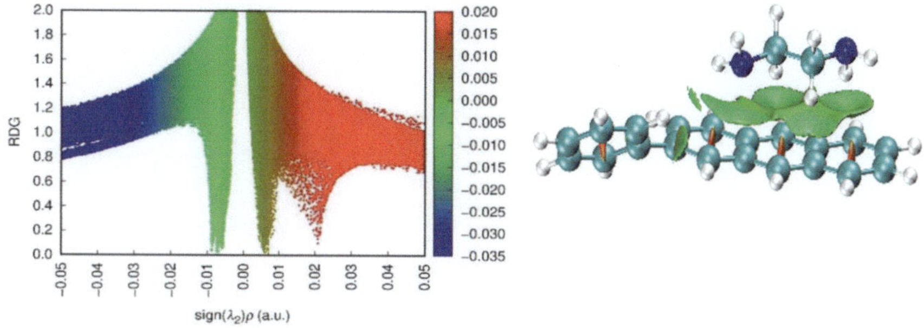

Fig. 3.29 Scatter plot and filled isosurfaces of the WYM molecular model with aromatic side chains vs. ethylenediamine extractant

on the aromatic side chain and the H atom on the ethylenediamine. However, by observing the scatter plot, it can be seen that the spike has a lower peak tip value and a smaller peak shape, so the intermolecular attraction caused by this weak hydrogen bond is smaller, and there is no other spike with a smaller sign(λ_2)ρ(a.u.) value other than this spike, suggesting that the spike with a sign(λ_2)ρ(a.u.) value of about −0.01 is the only weak mutual attraction between the two molecules other than the Van der Waals interaction. Therefore, the ethylenediamine extractant has a weak effect on the extraction of WYM reactive groups whose side chains are aromatic groups.

(b) **WYM Reactive Group and NMP Extractant**

Figure 3.30 shows the scatter plot and color-filled isosurface plots of the molecular model of WYM reactive moiety with –OH side chain and NMP extractant, from which it can be seen that a large number of mass-dense spikes exists on the scatter plot at sign(λ_2)ρ(a.u.) values between −0.005 and 0.005, which suggests that there are a large number of van der Waals interactions between the NMP molecule and the WYM molecule, whereas There is a weaker spike located at sign(λ_2)ρ(a.u.) of about −0.014, which suggests that there is a stronger than van der Waals interaction between the two molecules. Combined with the isosurface plots, it can be seen that there is a weak C–H–O hydrogen bond between the WYM molecule and the NMP molecule, and by comparing the spike peak tip values corresponding to this hydrogen bond on the scatter plot, it can be seen that the interaction force generated by this hydrogen bond is less than the force of the C–H–O hydrogen bond formed between the NMP molecule and the YM molecule in the previous subsection. Therefore, the NMP extractant also has a strong extraction ability for the WYM reactive group whose side chain is –OH, but this extraction ability is smaller than the extraction ability of the NMP extractant for the WYM reactive group whose side chain is a carbon–oxygen group.

Figure 3.31 shows the molecular model of WYM reactive moiety with carbon–oxygen groups in the side chain with NMP extractant scatterplot and color-filled isosurfaces, from which it can be seen that there exists an obvious spike at sign(λ_2)ρ(a.u.) value of about −0.025, which is not similar to that at sign(λ_2)ρ(a.u.)

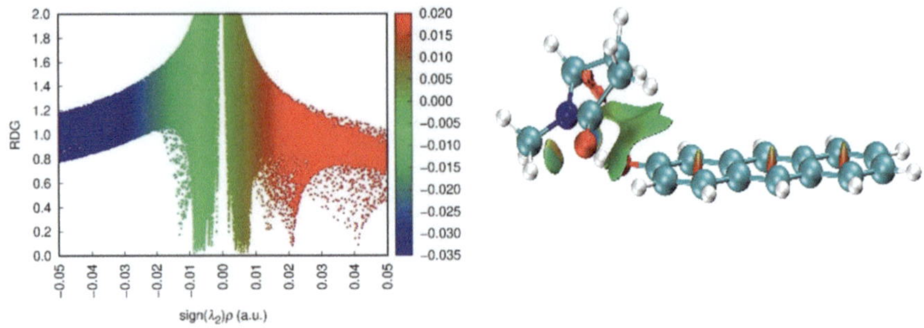

Fig. 3.30 Scatter plot and filled isosurfaces of WYM molecular model with –OH group in the side chain versus NMP extractant

value greater than 0 with a positive peak value of similar spike, indicating that the spike is a stronger hydrogen bond between the NMP and WYM molecules. Combined with the analysis of the isosurface diagram, it can be seen that the spike corresponds to the C–H–O hydrogen bond formed between the NMP molecule and the WYM molecule. From the relative positions of the two molecules, it can be seen that the interaction of the NMP molecule on the WYM molecule, whose side chain is a carbon–oxygen group, is mainly concentrated in the carbon–oxygen side chain, and part of the interaction force is located in the H–π bond formed between the aromatic ring of the WYM molecule and the NMP molecule. According to the interspike peak values and peak areas at the scatter plot sign(λ_2)ρ(a.u.) values less than 0, it can be seen that the NMP extractant has a strong extraction effect on the WYM reactive groups with carbon–oxygen groups in the side chains.

Figure 3.32 shows the scatter plots and color-filled isosurfaces of the molecular model of WYM reactive group with aliphatic side chain and the NMP extractant. It can be seen from the scatter plots that the interaction force generated by the NMP molecules on the WYM molecules with aliphatic side chains is very small compared with the interaction forces of the NMP molecules on the WYM molecules with –OH side chains and WYM molecules with carbon–oxygen moieties with carbon oxides,

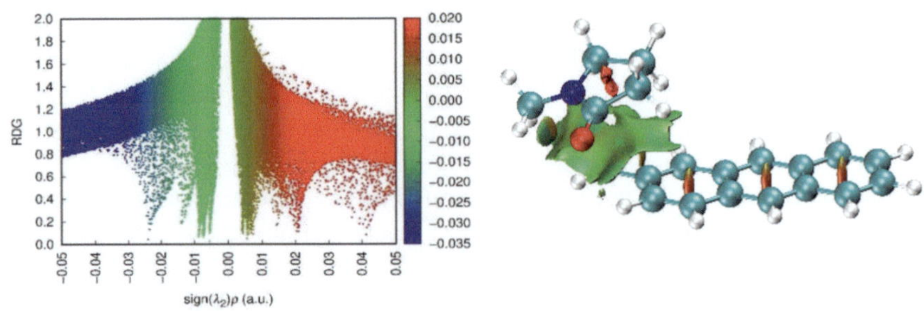

Fig. 3.31 Scatter plot and filled isosurfaces of WYM molecular model with carbon–oxygen groups in the side chain versus NMP extractant

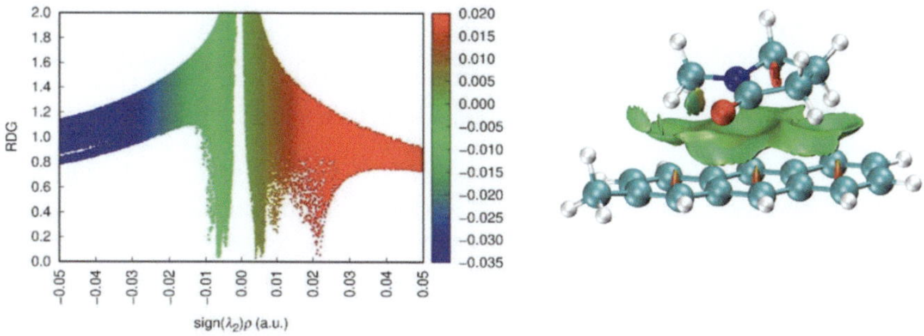

Fig. 3.32 Scatter plot and filled isosurfaces of WYM molecular model with aliphatic groups in the side chain versus NMP extractant

and there are only two stronger than van der Waals' forces on the WYM molecules with aliphatic side chains at the sign(λ_2)ρ(a.u.) values located at −0.01 and −0.008 respectively, there are two interaction forces stronger than the van der Waals forces. Combined with the analysis of the isosurface plots, it can be seen that these two spikes correspond to the hydrogen bond formed between the O atom and the H atom on the aliphatic side chain in ethylenediamine and the weak hydrogen bond formed between two H atoms, respectively. Although the hydrogen bond exists, the constraint of the relative spatial position makes the hydrogen bond exhibit weak mutual attraction, and comparing the spike at the sign(λ_2)ρ(a.u.) value <0 with the other spikes in the scatter plot also at the sign(λ_2)ρ(a.u.) value located in the value of the negative value, the NMP extractant is obviously weak for the WYM reactive group with aliphatic side chain. The extraction ability is weaker.

Figure 3.33 shows the molecular model of WYM reactive group with aromatic side chain and NMP extractant scatter plot and color-filled isosurface diagram, from the scatter plot, it can be seen that there are two obvious spikes at the sign(λ_2)ρ(a.u.) value of about ±0.005. Combined with the isosurface diagram, it can be seen that these two spikes correspond to the van der Waals forces between the WYM reactive group and the NMP molecule; The sign(λ_2)ρ(a.u.) value is about −0.015, which corresponds to a weak hydrogen bond between the N atom in the NMP molecule and the H atom on the aromatic ring of the WYM reactive group molecule in the isosurface diagram; and the spike at the sign(λ_2)ρ(a.u.) value of −0.01 corresponds to a weak hydrogen bond between the H atom of the NMP molecule and the H atom on the aromatic ring of the WYM reactive group molecule in the isosurface diagram; the spike at −0.01 corresponds to the weak hydrogen bond between the H atom of the WYM reactive group molecule and the H atom on the aromatic ring of the WYM molecule in the isosurface diagram. The weak hydrogen bond between the H atoms on the NMP molecule and the H atoms on the WYM molecule on the isosurface diagram. Analyzing the isosurface diagram as a whole, it can be seen that the interaction force between NMP and WYM molecules all acts on the benzene ring of NMP molecules and WYM active group molecules, and there does not exist any interaction with the group attached to the side chain, so the extraction effect of NMP extractant on WYM active group whose side chain is aromatic is generally effective.

(c) WYM Active Group and Petroleum Ether Extractant

Figure 3.34 shows the molecular model of WYM reactive group with –OH side chain and petroleum ether extractant scatter plot and color-filled isosurface diagram, from the scatter plot, it can be clearly seen that there is basically no intermolecular interaction force between WYM reactive group with –OH side chain and petroleum ether molecule except for van der Waals' force and weak hydrogen bonding at the value of sign(λ_2)ρ(a.u.) of about ±0.005, so the extraction ability of petroleum ether to WYM is general. Petroleum ether has a general ability to extract WYM reactive groups with –OH side chains.

Figure 3.35 shows the molecular model of the WYM reactive group with carbon–oxygen group in the side chain with the petroleum ether extractant scatter plot and the filled isosurface, from the scatter plot, it can be seen that there exists a spike with small density at the sign(λ_2)ρ(a.u.) value of about −0.012, which corresponds to the

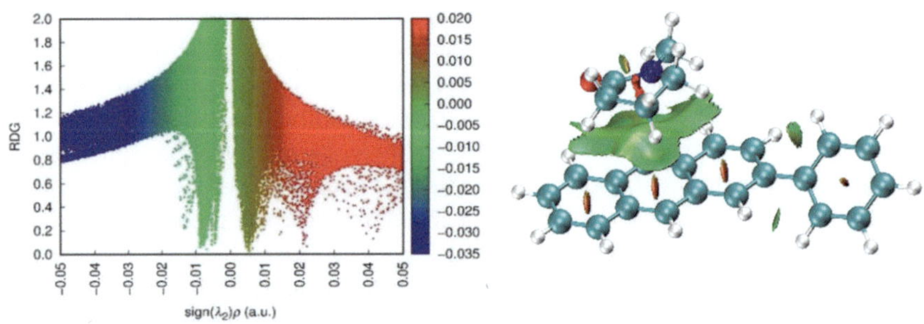

Fig. 3.33 Scatter plot and filled isosurfaces of WYM molecular model with aromatic side chains versus NMP extractant

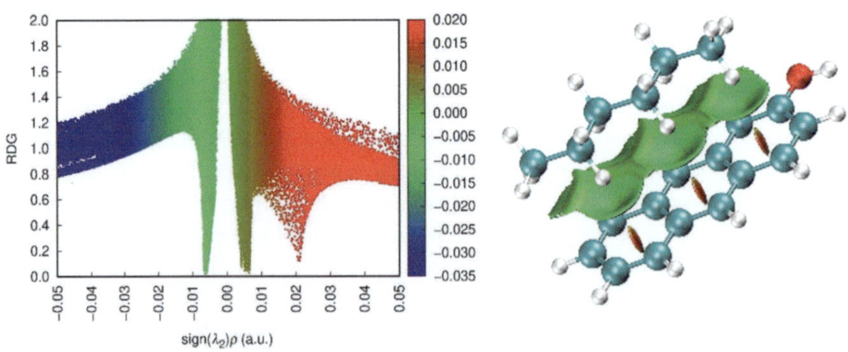

Fig. 3.34 WYM molecular model with hydroxyl energy group as side chain and scatter plot and color-filled isoplane plot of petroleum ether extractant

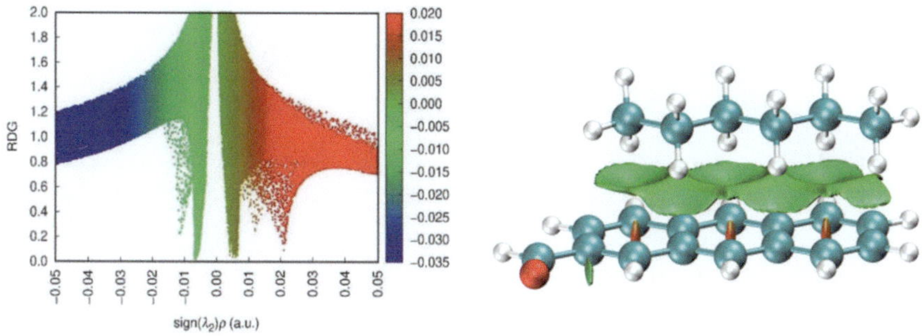

Fig. 3.35 Scatter plot and filled isosurfaces of WYM molecular model with petroleum ether extractant with carbon–oxygen groups in the side chain

formation of the H–π bond between the O atom on the side chain of the carbon–oxygen group and the H atom on the neighboring C atom on the isosurface, the interaction is weak and does not belong to the interaction of these two molecules to this WYM molecule, so there is only the interaction of these two molecules to this WYM molecule. H–π bond, which is a weak interaction and is not part of the interaction of the petroleum ether molecule on this WYM molecule, so there is only a van der Waals interaction between these two molecules at a sign(λ_2)ρ(a.u.) value of about ±0.005. Therefore, the petroleum ether extractant is generally effective in extracting WYM reactive groups with carbon–oxygen groups in the side chains.

Figure 3.36 shows the molecular model of WYM reactive group with aliphatic side chain with petroleum ether extractant scatter plot and color-filled isosurface map, from which it can be seen that the WYM reactive group with aliphatic side chain with petroleum ether extractant ADG scatter plot has a big difference with other petroleum ether extractant ADG scatter plot. A new spike appears at the sign(λ_2)ρ(a.u.) value of about −0.001, which corresponds to the dispersion force formed between the terminal H atom of the petroleum ether and the H atom on the aliphatic side chain on the isosurface map, so that removing the effect of van der Waals forces between the two molecules, there is also a weak interaction force between the petroleum ether and the aliphatic side chain. Therefore, petroleum ether has a certain extraction ability for WYM active groups whose side chains are aliphatic groups.

Figure 3.37 shows the scatter plot and color-filled isopotential surface diagram of the model of WYM reactive group molecule with aromatic side chain and petroleum ether extractant. From the scatter plot, it can be seen that there exists a spike at the sign(λ_2)ρ(a.u.) value of about −0.008, which, combined with the analysis of the isopotential surface diagram, corresponds to the weak hydrogen bond formed between the H atom on the benzene ring side chain of the WYM reactive group molecule and the petroleum ether-terminal H atoms of the WYM reactive group molecule and the H atom on the side chain of the benzene ring of the petroleum ether terminal; while the spike with a sign(λ_2)ρ(a.u.) value of about −0.006 corresponds to the weak hydrogen bond formed between the H atom on the approximate C atom of the WYM molecule and the H atom on the side chain of the aryl ring in the isotopic

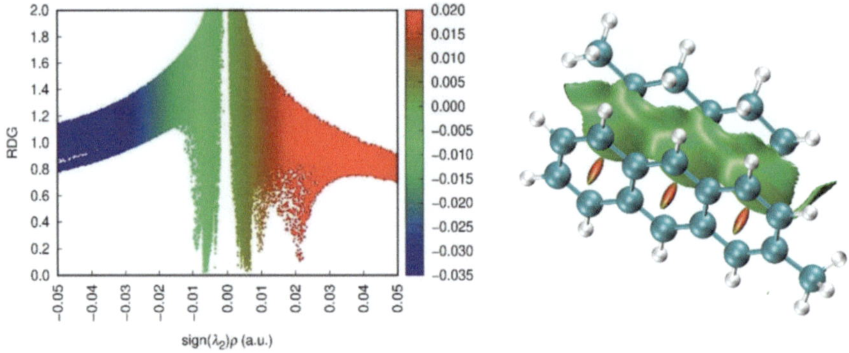

Fig. 3.36 Scatter plot and filled isosurfaces of WYM molecular model with petroleum ether extractant for aliphatic groups in the side chain

surface diagram, and the spike corresponds to the weak hydrogen bond between the H atom on the neighboring C atom and the H atom on the side chain of the aryl ring in the isotopic surface diagram. It does not affect the interaction of petroleum ether on the WYM reactive group. Therefore, the petroleum ether only has a weak extraction effect on the WYM reactive groups with aromatic side chains.

(d) **WYM Reactive Group and Toluene Extractant**

Figure 3.38 shows the scatter plot of WYM molecules with –OH side chain and toluene extractant and the filled isosurface plots. The ADG scatter plot of WYM molecules with –OH side chain and toluene extractant is similar to that of WYM reactive groups with carbon–oxygen groups and toluene extractant in the previous subsection, and the combination of isosurface plots reveals that this is because the toluene molecule likewise forms weak interactions with the WYM molecules only with the aromatic group, and the change in the side chain has no effect on this weak interaction.

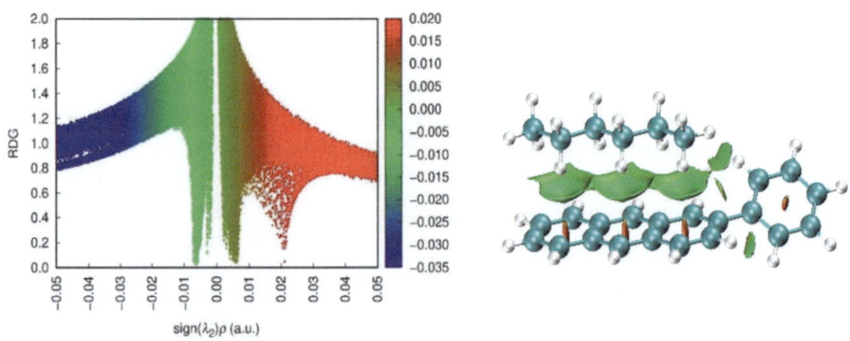

Fig. 3.37 Scatter plot and filled isosurfaces of WYM molecular model with aromatic side chains versus petroleum ether extractant

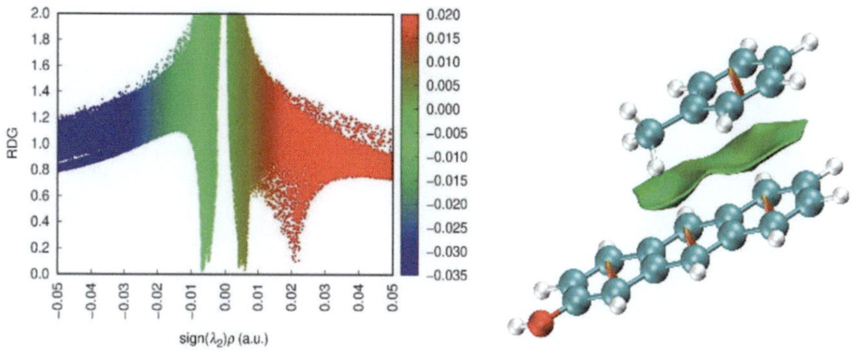

Fig. 3.38 Scatter plot and filled isosurfaces of WYM molecular model with toluene extractant with –OH group in the side chain

Figure 3.29 shows the molecular model of the WYM reactive moiety with aromatic side chain as an aromatic moiety with the scatter plot and color-filled isosurface plots of the ethylenediamine extractant. By observing the scatter plots, it can be seen that the spikes with negative sign(λ_2)ρ(a.u.) values are mainly concentrated in the range of −0.005 to −0.01, and the spike with the sign(λ_2)ρ(a.u.) value of about −0.01 suggests that the Ethylenediamine molecule and the side chain is an aromatic group there may be a weak interaction force, combined with the isosurface plot it can be seen that the spike corresponds to the hydrogen bond formed between the H atom on the aromatic side chain and the H atom on the ethylenediamine. However, by observing the scatter plot, it can be seen that the spike has a lower peak tip value and a smaller peak shape, so the intermolecular attraction caused by this weak hydrogen bond is smaller, and there is no other spike with a smaller sign(λ_2)ρ(a.u.) value other than this spike, suggesting that the spike with a sign(λ_2)ρ(a.u.) value of about −0.01 is the only weak mutual attraction between the two molecules other than the Van der Waals interaction. Therefore, the ethylenediamine extractant has a weak effect on the extraction of WYM reactive groups whose side chains are aromatic groups (Fig. 3.39).

Figure 3.40 shows the scatter plot and color-filled isosurface plots of WYM molecule with aliphatic group in the side chain and toluene extractant. From the scatter plots, it can be seen that compared with other ADG scatter plots of toluene extractant, there are several smaller spikes in the sign(λ_2)ρ(a.u.) value of this plot at the range of −0.005 to −0.005, and combined with the analysis of the isosurface plots, it can be seen that these spikes correspond to the weak hydrogen bonding between the aryl ring of the toluene and the WYM molecule aryl ring, from the sign(λ_2)ρ(a.u.) negative value region, the spike at this place compared with the ADG scatter plot of the WYM molecular model with methyl and carbon–oxygen group in the side chain and the WYM molecular model with –OH in the side chain showed a larger peak tip value and a larger number of spikes, but the majority of them were van der Waals interactions, so it was judged that toluene had a strong influence on

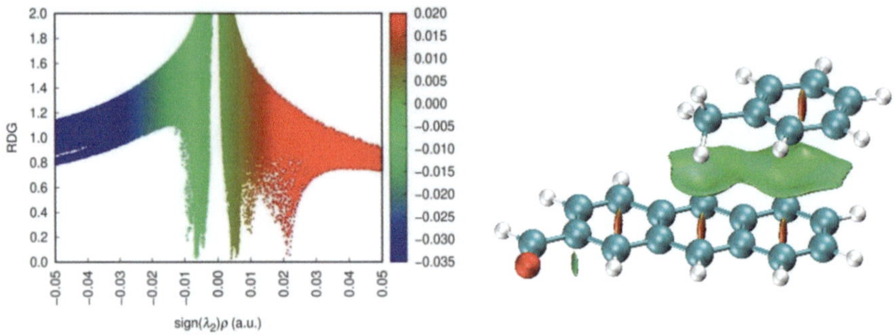

Fig. 3.39 Scatter plot and filled isosurface of molecular model of WYM reactive moiety with carbon–oxygen groups in the side chain versus toluene extractant

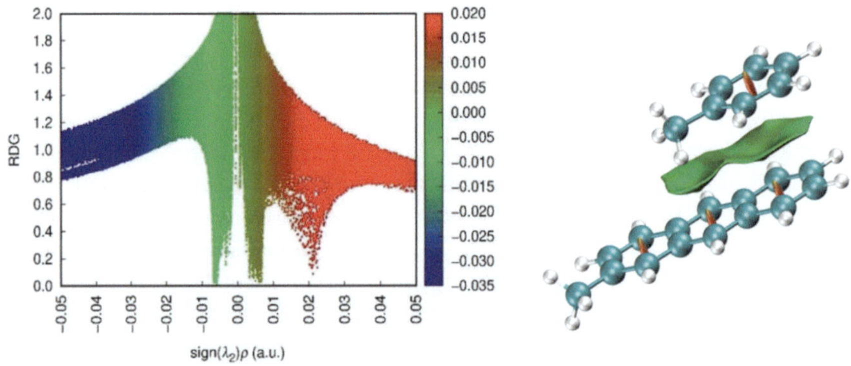

Fig. 3.40 Scatter plot and filled isosurfaces of WYM molecular model with aliphatic groups in the side chains versus toluene extractant

the activity of the WYM group molecule aliphatic side chain has a weaker extraction effect.

Figure 3.41 shows the molecular model of WYM reactive group with aromatic side chain and toluene extractant scatter plot and color-filled isosurface plots, there are four spike values at the sign(λ_2)ρ(a.u.) value >0 in the scatter plot, and excluding the van der Waals interactions corresponding to the sign(λ_2)ρ(a.u.) value of 0.005, the remaining three spikes correspond to the isosurface plots respectively face diagrams of the site-barrier effect in the aromatic ring of the WYM molecule, the site-barrier effect in the toluene aromatic ring, and the site-barrier effect in the aromatic side chain; and the spike with a sign(λ_2)ρ(a.u.) value of −0.012 is the weak hydrogen-bonding interactions between the neighboring H atoms on the aromatic side chain and the H atoms on the aromatic ring of the WYM reactive group molecule. The weak interaction forces corresponding to these spikes are all interactions formed between the NMP molecules and the WYM active group molecules themselves, and it is difficult to judge the extraction effect of toluene on the WYM active group molecules. However, it can be seen through the analysis of the isosurface diagrams that the toluene molecules and the aromatics on the WYM form spatial stacking on top of each other due to the formation of π–π bonding between the benzene ring in the toluene molecule and the benzene ring in the WYM molecule. This is due to the formation of π–π bond between the benzene ring in the toluene molecule and the benzene ring in the WYM molecule, which is the most attractive bond in the weak interactions, and the existence of π–π bond makes the toluene molecule and the WYM molecule form an exceedingly stable spatial structure, and therefore, it is believed that the toluene molecule has a very strong extraction ability for the aromatic substances.

The mode of interaction between four different side chain WYM reactive groups and four different extractant molecules is shown in Table 3.3.

As can be seen from the table, the weak interaction forces of the extractants on the coal molecules are mainly manifested as hydrogen bonding, weak hydrogen bonding, and van der Waals forces of different strengths, in which NMP has a stronger

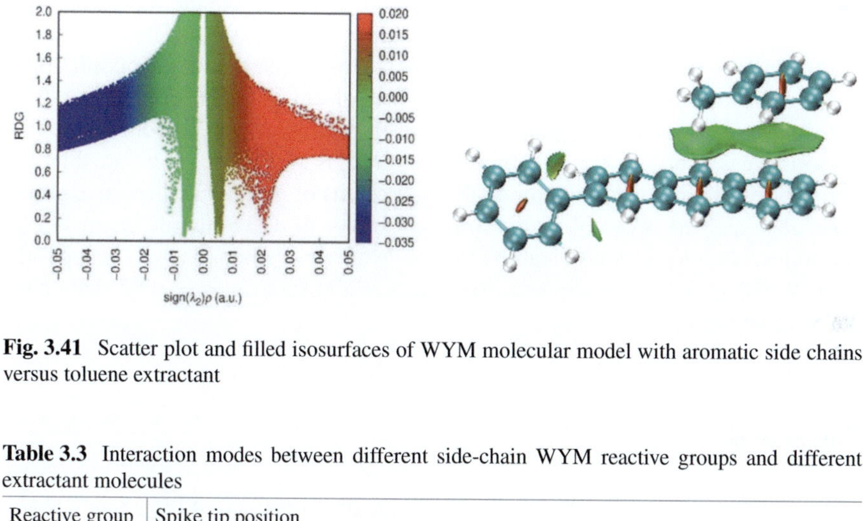

Fig. 3.41 Scatter plot and filled isosurfaces of WYM molecular model with aromatic side chains versus toluene extractant

Table 3.3 Interaction modes between different side-chain WYM reactive groups and different extractant molecules

Reactive group	Spike tip position			
	Ethylenediamine	NMP	Petroleum ether	Toluene
–OH	−0.0012; −0.008	−0.004; −0.009; −0.013	−0.007	−0.007
	0.005; 0.13	0.005; 0.014	0.005	0.005
–CHO	−0.005	−0.006; −0.009; −0.014	−0.007; −0.011	−0.006; −0.01
	0.005; 0.01	0.005; 0.015	0.005; 0.011	0.005; 0.01
$–CH_3$	−0.006	−0.007; −0.009	−0.005; −0.009	−0.006
	0.005; 0.009	0.005; 0.01	0.005; 0.01	0.005
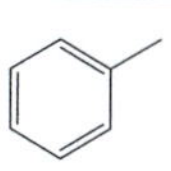	−0.012	−0.005; −0.009; −0.013	−0.006	−0.007; −0.011
	0.03	0.005; 0.015	0.005	0.005; 0.012

extraction effect on the model of coal molecules carrying the –OH side chain or carbon–oxygen group side chain in the WYM molecules; the toluene extractant is more targeted at the aromatic group in the WYM, while the petroleum ether has a better extraction effect on the aliphatic group side chain. By comparing the interaction forces and spike tip positions of coal molecules carrying different side chains with different extractants, it can be found that NMP molecules can form hydrogen bonds with WYM molecules with –OH side chains, while weak interaction forces are formed with coal molecule models with other side chains. The corresponding spike on the scatter plot, both in number and size, is smaller than that corresponding to the interaction with –OH group, and the spike peak tip position is closer to 0.

Its extraction ability for WYM molecules containing different side chains is ranked as –OH side chain WYM molecules > carbon–oxygen group side chain WYM molecules > aromatic side chain WYM molecules > aliphatic side chain WYM molecules. NMP can form more weak hydrogen bonds with coal molecule model with –CHO side chain, which corresponds to more and denser spikes for mutual attraction interactions. Petroleum ethers form weak hydrogen bonds with toluene and aliphatic side chains of WYM reactive group molecules, but the weak hydrogen bonds with aliphatic group correspond to larger spike peaks, and their extraction ability for WYM molecules with different side chains is ranked as aliphatic side chain WYM molecules > aromatic side chain WYM molecules > –OH side chain WYM molecules > carbon–oxygen group side chain WYM molecules. Toluene molecules can form weak hydrogen bonds with WYM molecules whose side chains are toluene, while there is no mutual attraction to the side chains of other WYM reactive group molecules, and the order of their extraction ability to different side chain WYM molecules is aromatic side chain WYM molecules > carbon–oxygen group side chain WYM molecules > aliphatic side chain WYM molecules > –OH side chain WYM molecules.

References

Çağlayan G, Ipek U (2021) Desulfurization of coal with oak ash extraction. Energy Sour A Recov Utiliz Environ Effects 12:1–10. https://doi.org/10.1080/15567036.2021.1920648

Chen C, Kacarab M, Tang P et al (2016) SOA formation from naphthalene, 1-methylnaphthalene, and 2-methylnaphthalene photooxidation. Atmosp Environ 131:424–433. https://doi.org/10.1016/j.atmosenv.2016.02.007

Frisch M, Trucks G, Schlegel H et al (2013) Gaussian 09 (Revision D.01), Gaussian

Gao ZY, Yang WJ (2017) Adsorption mechanism of water molecule on different rank coals molecular surface. J China Coal Soc 42(03):753–759

Gonzalez C, Schlegel H (1989) An improved algorithm for reaction path following. J Chem Phys 90(4):2154–2161. https://doi.org/10.1063/1.456010

Grimme S, Antony J, Ehrlich S et al (2010) A consistent and accurate ab initio parametrization of density functional dispersion correction (DFT-D) for the 94 elements H-Pu. J Chem Phys 132:154104. https://doi.org/10.1063/1.3382344

Jiang YX (2023) Research on hydrogen bonding interactions between allyl/ester-functionalized ionic liquids and DMSO. Qingdao. https://doi.org/10.27262/d.cnki.gqdau.2023.001086

Kato S, Kuwako S, Takahashi N et al (2016) Benzo-and naphthopentalenes: syntheses, structures, and properties. J Organ Chem 81(17):7700–7710. https://doi.org/10.1021/acs.joc.6b01409
Li Z, Zheng Z, Li X et al (2019) Enhancing metal separations using hydrophilic ionic liquids and analogues as complexing agents in the more polar phase of liquid–liquid extraction systems. Ind Eng Chem Res 58(34):15628–15636. https://doi.org/10.1021/acs.iecr.9b03472
Li CH, He H, He MJ et al (2022) Extraction kinetics of Ce(IV) from nitric acid solutions using ionic liquid. CIESC J 73(04):1606–1614
Liu X, Hu Z, Hu D et al (2024) Degradative solvent extraction of low rank coal: structural evolution of extraction products. J Anal Appl Pyrol 177:106290. https://doi.org/10.1016/j.jaap.2023.106290
Liu C (2022) Research on the interaction mechanism during medium- and low temperature coal tar DMSO extraction proceedings. China University of Mining and Technology, Beijing. https://doi.org/10.27624/d.cnki.gzkbu.2022.000054
Lu Y, Li E, Cheng H et al (2022) Effect of oxygen functional groups on the surface properties and flotation response of fine coal, comparison of rank with oxidation. Int J Coal Preparat Utiliz 80:1469. https://doi.org/10.1080/19392699.2018.1471469
Luo Q, Wen L, Qi Z (2023) The mechanism of ozone oxidation of coal and the revelation of coal macromolecular structure by oxidation products. ACS Omega 9(1):753–770. https://doi.org/10.1021/acsomega.3c06525
Ma X, Dong X, Fan Y (2018) Prediction and characterization of the microcrystal structures of coal with molecular simulation. Energy Fuels 32(3):3097–3107. https://doi.org/10.1021/acs.energyfuels.7b03698
Mao KM, Mo WL, Ma FY et al (2021) Composition and structure characteristics of soluble organic matter from Naomaohu lignite by sequential extraction and thermal conversion performance of the corresponding residue. J Fuel Chem Technol 49(10):1389–1401
Sangeetha T, Sahana R, Mounica P et al (2023) Atoms in molecules theory, electrostatic potential surface and frontier molecular orbital analyses on water multimers and pyridine: water hydrogen bonded complexes. Comput Theor Chem 1219:113960. https://doi.org/10.1016/j.comptc.2022.113960
Shiga T, Kumamaru R, Newton G et al (2020) Heteroleptic iron (ii) complexes with naphthoquinone-type ligands. Dalton Trans 49(5):1485–1491. https://doi.org/10.1039/C9DT03946E
Wang WT, Su Z, Xie SB et al (2021) Research progress on the microstructure of the third phase during the liquid-liquid extraction of spent nuclear fuel reprocessing. J Nucl Radiochem 43(01):39–49
Wang JR, Jin ZX, Deng CB (2007) Quantum chemical theory of coal self-ignition. Beijing
Wang ZY (2023) Analysis of the properties of clusters ConMoS(n=1−5) based on density functional theory, Lingning. https://doi.org/10.26923/d.cnki.gasgc.2023.000786
Xu CM, Yin MF, Tong H et al (2023) Theoretical calculation and experimental verification of aromatics extraction from naphtha by ionic liquids. J China Univ Petrol 47(05):138–145. https://doi.org/10.3969/j.issn.1673-5005.2023.05.0014
Ye YX, Ding XQ, Chi HR et al (2022) Hydrophobic deep eutectic solvent hydrogen bond interaction regulation and extraction performance of copper. Chem Ind Eng Prog 41(S1):397–406
Zhang YC, Hao CY, He WH et al (2021) Quantum chemistry study on effects of NH_3 on physical oxygen adsorption of coal. Coal Conver 44(06):16–25
Zhang ZH, Xiao Q, Lin H et al (2023) Determination of cationic surface active bactericides in disinfectants based on ion exchange solid phase extraction-high performance liquid chromatography-tandem mass spectrometry. J Anal Sci 39(05):537–542
Zhou BZ (2022) Characteristics evolution during low temperature oxidation of coal influenced by multi-component gases competitive adsorption. Xuzhou. https://doi.org/10.27623/d.cnki.gzkyu.2022.002755

Chapter 4
Ultrasonic Extraction of Coal Reactive Groups and Microstructure Evolution Pattern

Abstract The technology of solvent extractions assisted by ultrasonic was applied to extract active groups in coal molecules. Then, changes of the active groups before and after the extraction of coal samples and their evolution laws in the process of oxidation heating were analyzed symmetrically by employing electron paramagnetic resonance spectroscopy (EPR) and Fourier transform infrared spectroscopy (FT-IR). Consequently, extraction effects and mechanisms of different extractants on the active groups in coal were revealed. At the same time, mechanisms of weak interaction damage of coal molecular structure in the extraction process were revealed in this Chapter, which provided a new idea and an important basis for the study of coal microstructure and oxidation characteristics.

Extraction (Guo et al. 2023) is one of the effective methods to study the chemical structure of coal and the role of key groups in its oxidation process, the essence of which is to separate individual groups in coal from the coal molecular network through the interaction of the extraction solvent with the structure of coal macromolecules (Xue et al. 2022). Ultrasonic extraction, on the other hand, introduces the cavitation effect of ultrasound and the additional turbulence, perturbation, interfacial and aggregation effects generated by ultrasound for solvent extraction to improve the extraction efficiency.

In this chapter, on the basis of the molecular structure of the coal samples obtained in Chap. 3, combined with the four organic solvents preferred in Chap. 3, ultrasonic extraction was carried out on the coal samples. After the extraction, the changes of free radicals and functional groups in coal before and after extraction were analysed using electron paramagnetic resonance spectrometry (EPR) and Fourier transform infrared spectroscopy (FT-IR), and then combined with in-situ infrared spectroscopy (In-situ FT-IR), the changes of microstructural variation rules in the oxidation process of the coal samples before and after extraction were analysed, which provided a basis for the subsequent determination of the coal auto-ignition reactive groups in the Its lays the foundation for the subsequent determination of the oxidation reaction course of the active groups of coal auto-ignition and revealing its gas generation mechanism.

J. Deng et al., *Coal Spontaneous Combustion Theory with Chain Self-promoted Oxidizing Induced by Active Group*, https://doi.org/10.1007/978-981-96-5362-1_4

4.1 Ultrasonic Extraction of Coal Reactive Groups

Owing to the extremely complex molecular structure of coal, it is not possible to directly study the macromolecules of coal samples. Based upon the analysis of the weak interactions between the reactive groups of coal samples and organic solvents in Chap. 3, some of the groups in coal were extracted from the macromolecular system by using the preferred four organic solvents ethylenediamine, NMP, petroleum ether, and toluene under ultrasonication assistance, thus realising the discretisation of the reactive groups in coal, and laying a foundation for the study of each reactive group-s oxidation reaction mechanism of each reactive group.

4.1.1 Experimental Setup and Test Methods

In this paper, ultrasonic extraction extraction of coal samples was carried out using KQ-100B ultrasonic cleaner under the action of different extraction solvents, and the experimental setup is shown in Fig. 4.1.

For ultrasonic extraction experiments (Dai et al. 2022), 10 g of coal samples were weighed into a round-bottomed flask each time, 500 mL of extraction solvent was added to the flask, and the flask was tightly capped and placed in an ultrasonic cleaner of type KQ-100B for extraction at room temperature for 4 h. At the end of the extraction, the extracted coal samples on the funnel were collected, and after rinsing with ethanol until no extract was detected (Liu 2022), the samples were put into a vacuum drying oven for drying, and then weighed after completion of the drying process. After drying, the coal samples were weighed and ground to 200 mesh, and the ground coal samples were put into wide-mouth bottles for sealing.

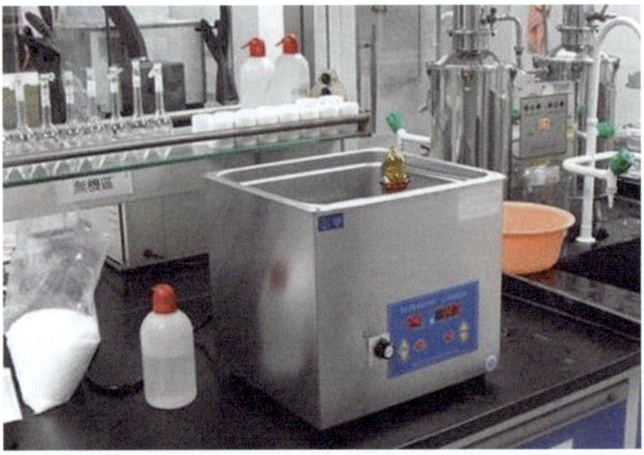

Fig. 4.1 KQ-100B ultrasonic cleaner

4.1.2 Analysis of Extraction Effect

To be able to express the ultrasonic extraction effect of coal samples concretely, the concept of extraction rate is introduced, and the calculation formula (Eq. 4.1) is as follows:

$$W_t = \frac{(1 - M_{ad}) \times M_1 - M_2}{(1 - M_{ad} - A_{ad}) \times M_1} \times 100\% \tag{4.1}$$

where, W_t—Extraction rate, %; M_{ad}—Water content of a coal sample; A_{ad}—Ash content of coal samples; M_1—Raw coal quality, g; M_2—Extracted coal sample quality, g;

Four ultrasonic extraction experiments were carried out using different organic solvents on coal samples under the same extraction conditions (Guo 2022). After the extraction experiments, the extracted coal samples were collected into a vacuum drying oven and weighed after drying, and the extraction rate was calculated from (Eq. 4.1), and the extraction rates of different extractants were shown in Tables 4.1 and 4.2.

Table 4.1 Extraction rates of different extractants for YM

Extraction agent	Raw coal quality/(g)	Extracted coal sample quality/(g)	Extraction percentage/(%)
Ethylenediamine	10.12	8.77	13.35
NMP	10.04	8.79	12.69
Petroleum ether	10.08	8.87	11.62
Toluene	10.07	9.02	10.47

Table 4.2 Extraction rate of WYM with different extractants

Extraction agent	Raw coal quality/(g)	Extracted coal sample quality/(g)	Extraction rate/(%)
Ethylenediamine	10.13	8.98	11.28
NMP	10.07	9.12	8.49
Petroleum ether	10.02	9.23	9.43
Toluene	10.12	9.26	7.83

4.2 Characteristics of Surface Group Changes of Coal Samples Before and After Extraction

4.2.1 Characteristics of Surface Group Changes Before and After YM Extraction

Figure 4.2 shows the infrared absorption spectra of coal samples after YM was extracted by four different extractants: Ethylenediamine, NMP, petroleum ether, and toluene. To further clarify the changes in the content of different types of groups in the residual coal samples after YM extraction by different extractants, the spectra in Fig. 4.2 are attributed to hydroxyl (–OH) groups, carbon–oxygen groups, aliphatic groups, and aromatic groups in this paper in accordance with the results of the peaks of the infrared spectroscopic analysis in Chap. 2. For this reason, the infrared absorption spectra of the extracted coal samples were fitted with waveband peaks based on Fig. 4.2.

(a) **Analysis of the proportion of hydroxyl group (–OH) in YM extracted coal samples**

The absorption peaks located between 3000 and 3700 cm^{-1} are mainly attributed to –OH, and Fig. 4.3 shows the peak fitting results of the absorption peaks between 3000 and 3700 cm^{-1} for YM extracted by different extractants. The integration of the peak splitting curves in Fig. 4.3 was used to obtain the content of –OH groups in the four extracted coal samples, and then the comparison of the content of –OH groups between YM and its four extracted coal samples was plotted, and the results are shown in Fig. 4.4.

As shown in Fig. 4.4, after extraction with different extractants, the percentage of –OH groups in the four extracted coal samples, from high to low, was toluene extracted coal sample (8.625%) > petroleum ether extracted coal sample (8.465%) > NMP extracted coal sample (8.427%) > ethylenediamine extracted coal sample (6.892%). That is to say, among the four extractants, ethylenediamine has the strongest ability to extract the –OH group in the YM molecule, and toluene has the weakest extraction ability.

The residual –OH group of toluene extracted coal samples is the highest among the four groups of samples, with a decrease of only 9.724%, which is due to the fact that toluene is a non-polar organic solvent, which doesn't contain atoms with strong electronegativity, such as O, N, F, etc., and does not have hydrogen bonding acceptor so that it cannot form hydrogen bonding with O atoms with strong electronegativity in the –OH group, and due to the difference of the polarities of organic solvents and target groups, toluene can't dissolve –OH O group effectively, so toluene has a poorer extraction effect on –OH. Because of the difference in polarity between the organic solvent and the target group, toluene cannot effectively dissolve the –OH group, so the extraction effect of toluene on –OH is poor, and the percentage of ethylenediamine-extracted coal residues of –OH group is the lowest, 6.892%, which is a decrease of

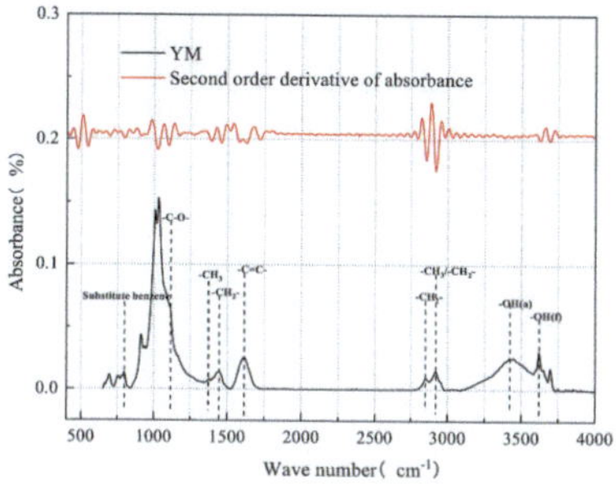

(a)YM infrared absorption spectrum

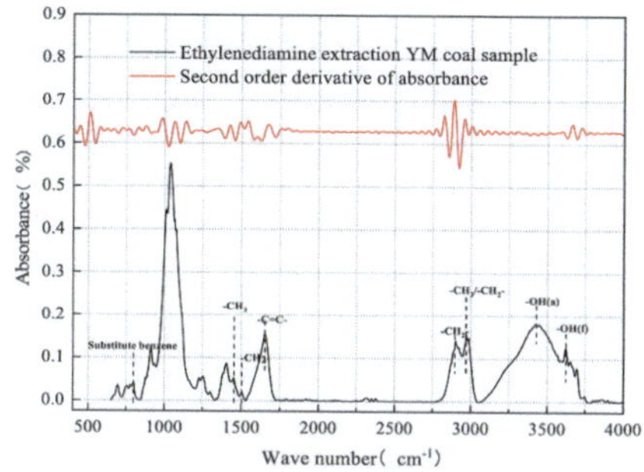

(b)Infrared absorption spectra of ethylenediamine extract YM coal samples

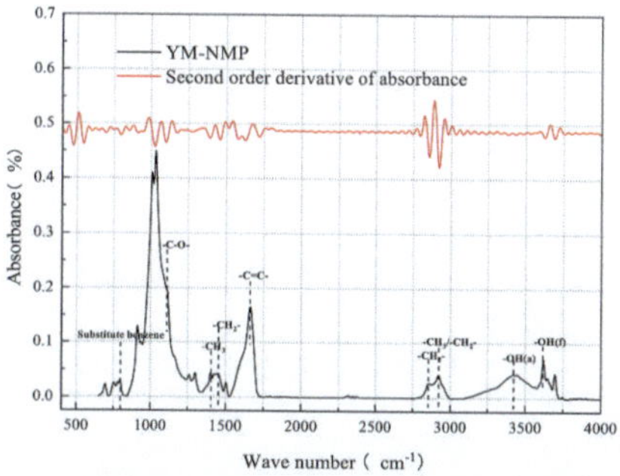

(c)Infrared absorption spectra of NMP extracted coal samples

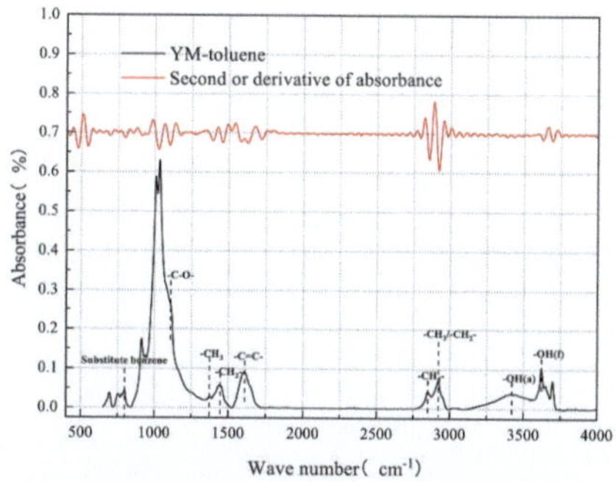

(d)Infrared absorption spectra of toluene YM extracted YM coal samples

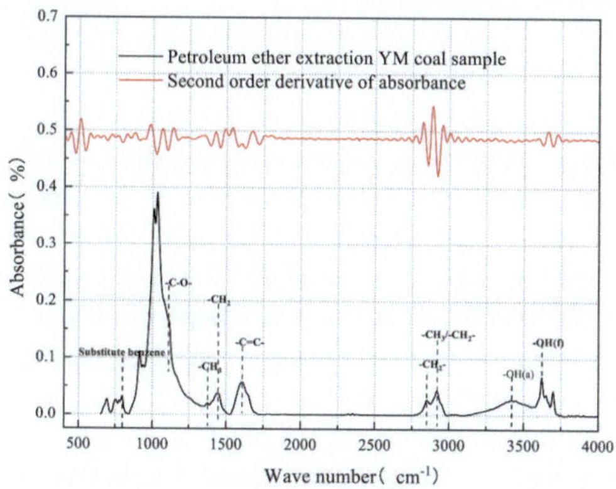

(e)Infrared absorption spectra of petroleum ether extraction YM coal samples

Fig. 4.2 Infrared absorption spectra of YM and its extracted coal samples analysis of the proportion of hydroxyl group (–OH) in YM extracted coal samples

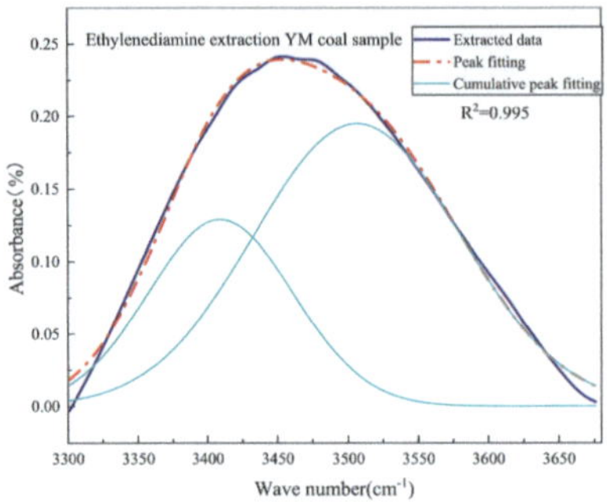

(a) ethylenediamine extraction YM coal samples –OH group content

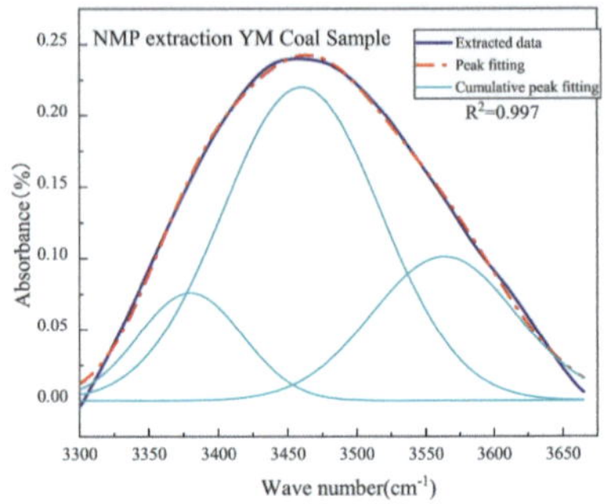

(b) NMP extracted YM coal samples –OH group content

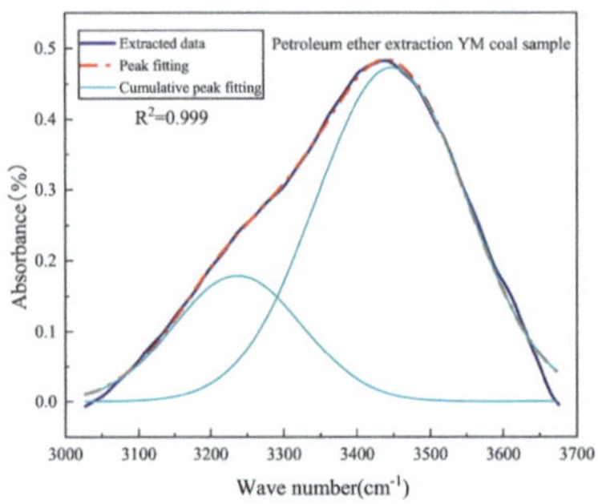

(c)Petroleum ether extraction YM coal samples –OH group content

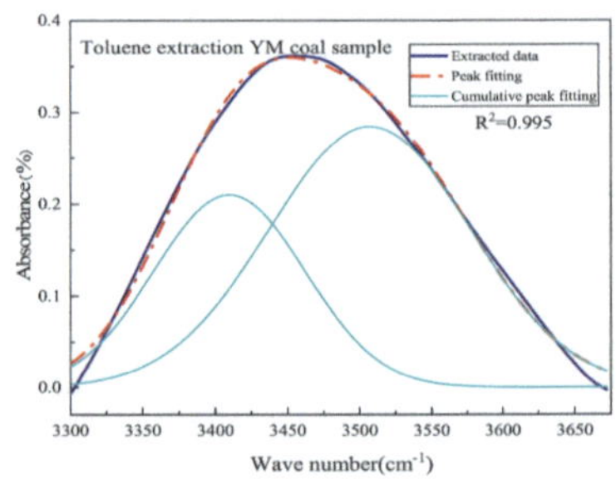

(d)Toluene extraction YM coal samples –OH group content

Fig. 4.3 Infrared spectra of hydroxyl (–OH) groups of four extracted coal samples

9.686% compared with YM, which is due to the fact that the N atoms in the two amino groups in the molecule of ethylenediamine are strongly electronegative, and they can become the acceptor of hydrogen bonding, thus forming a hydrogen bond with the H atoms in the –OH, which is similar to that in the weak interaction analysis in Chap. 3. This inference is consistent with the results of the isosurface plot analysis of the interaction between ethylenediamine and the coal molecule model with –OH as the side chain in the interaction analysis in Chap. 3.

The mutual attraction produced by hydrogen bonding is much stronger than other weak interactions formed between –OH and other extractants, and thus ethylenediamine is more effective in the extraction of –OH. In addition, by observing the content of different forms of –OH, it can be found that, unlike other groups, the loss of the three different forms of –OH has a large difference, and the maximum loss of the –OH can be up to 62.038% of the YM, compared with the phenol hydroxyl group, the maximum decrease is only 41.889%, which is due to the fact that the –OH mainly exists in the water molecules attached to the coal, and the extracted coal sample This is due to the fact that –OH mainly exists in the water molecules attached to the coal, and the extracted coal samples were evaporated by ethanol filtration, and

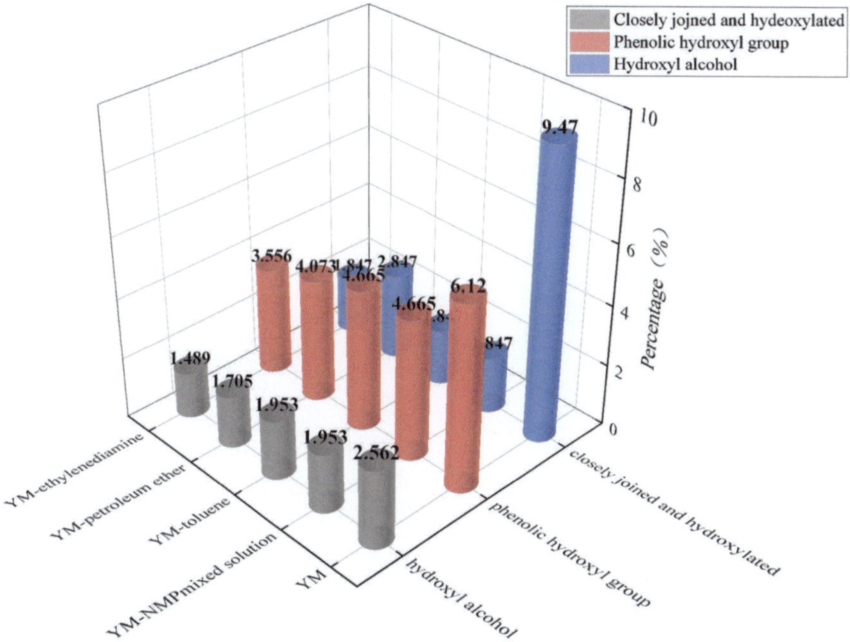

Fig. 4.4 Comparison of –OH group content between YM and its four extracted coal samples

the residual water decreased greatly, which was manifested as a substcnticl decrease in the content of conjugated –OH in the microcosm, so in fact, the extraction ability of ethylenediamine on the –OH group in the YM is mainly manifested in the extraction of phenolic hydroxyl and alcohol hydroxyl groups.

(b) **Analysis of the proportion of carbon and oxygen groups in YM extracted coal samples**

The absorption peaks located between 1500 and 1750 cm^{-1} were mainly attributed to carbon–oxygen groups. Figure 4.5 show the results of peak fitting of the absorption peaks between 1500 and 1750 cm^{-1} for coal samples after YM extraction by different extractants. Integrating the peak splitting curves in Fig. 4.5, the contents of carbon and oxygen groups in the four extracted coal samples were obtained, and then the carbon and oxygen group contents of YM and its four extracted coal samples were plotted, and the results are shown in Fig. 4.6.

By integrating the peaks of the fitted peaks, the contents of carbon and oxygen groups in the four extracted coal samples were obtained, and the carbon and oxygen group contents of YM and its four extracted coal samples were plotted, as shown in Fig. 4.6. As can be seen from the graphs, after extraction with different extractants, the percentage of carbon and oxygen groups in the four extracted coal samples, in descending order, was petroleum ether extracted coal sample (36.098%) > toluene extracted coal sample (35.681%) > ethylenediamine extracted coal sample (32.218%)

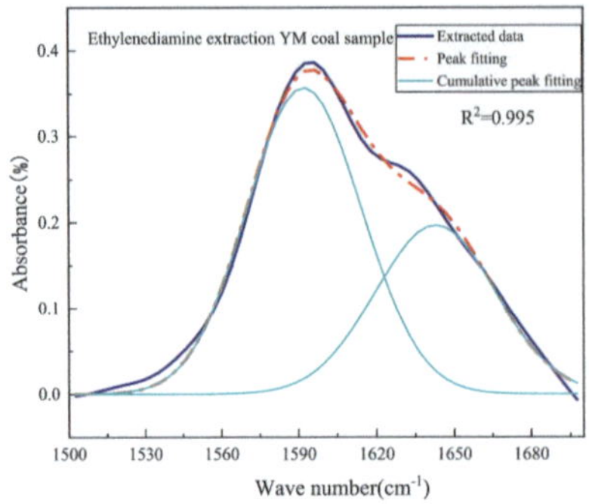

(a)Ethylenediamine extract YM Coal sample carbon and oxygen group content

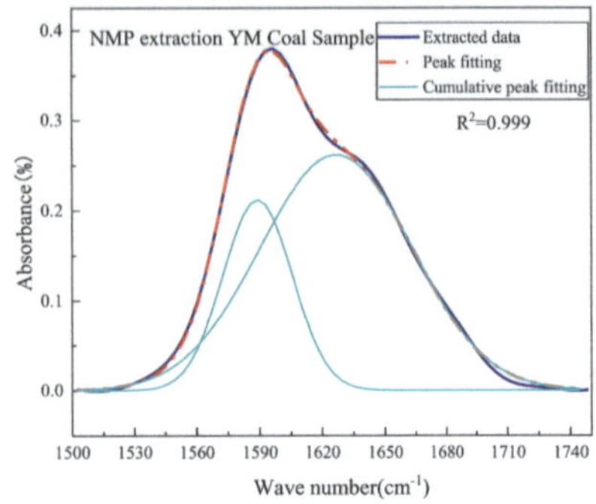

(b) Carbon-oxygen group content of NMP extracted YM coal samples

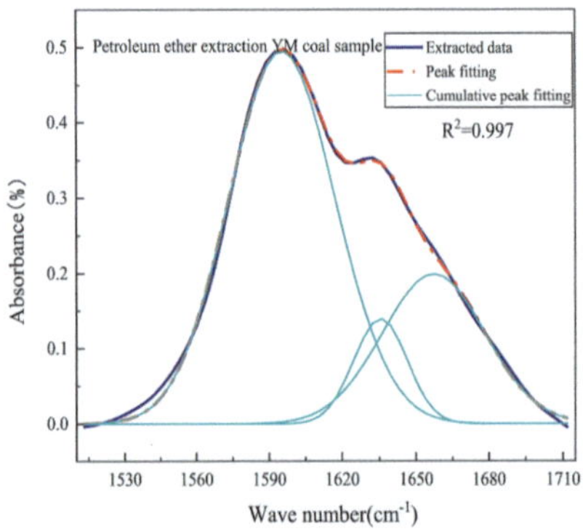

(c) Carbon-oxygen group content of petroleum ether extracted YM coal samples

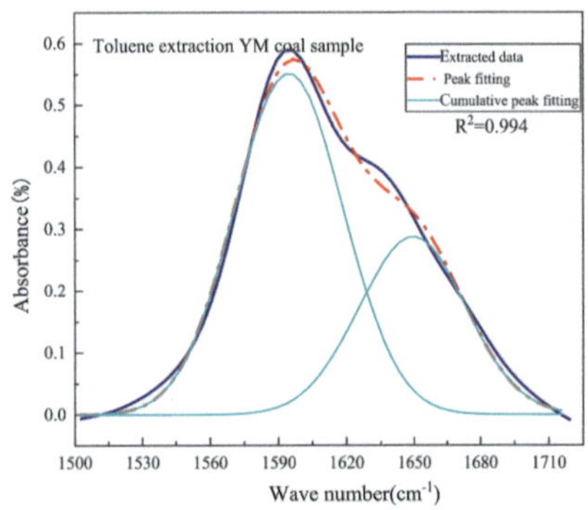

(d) Carbon-oxygen group content of toluene extracted YM coal samples

Fig. 4.5 Fitted peaks of carbon and oxygen groups of four extracted coal samples

> NMP extracted coal sample (31.045%). Therefore, among the four selected extractants, NMP has the strongest extraction ability for the carbon–oxygen groups in the YM molecule, while petroleum ether has the weakest extraction ability.

The percentage of residual carbon and oxygen groups of petroleum ether extracted coal samples is the highest among the four groups of samples, 35.098%, which is ca. 0.7% lower than that of the original coal of YM. According to the previous research on organic matter extraction and combined with the weak interaction analysis in Chap. 3, it can be seen that the petroleum ether selected in this paper is mainly composed of n-hexane, which, as a typical nonpolar substance, does not have any strong electronegative atoms, and cannot form typical N–H–O hydrogen bonds with the O atoms of the coal molecule (N is a strong electronegative element, such as N, F and O). It cannot form typical N–H–O hydrogen bond with O atom in C oxygen group of coal molecule (N is electronegative element, such as N, F, O, etc.), in short, the petroleum ether may not be able to effectively adsorb the molecular model of coal molecule whose side-chain is the carbonyl oxygen group, and form a stable

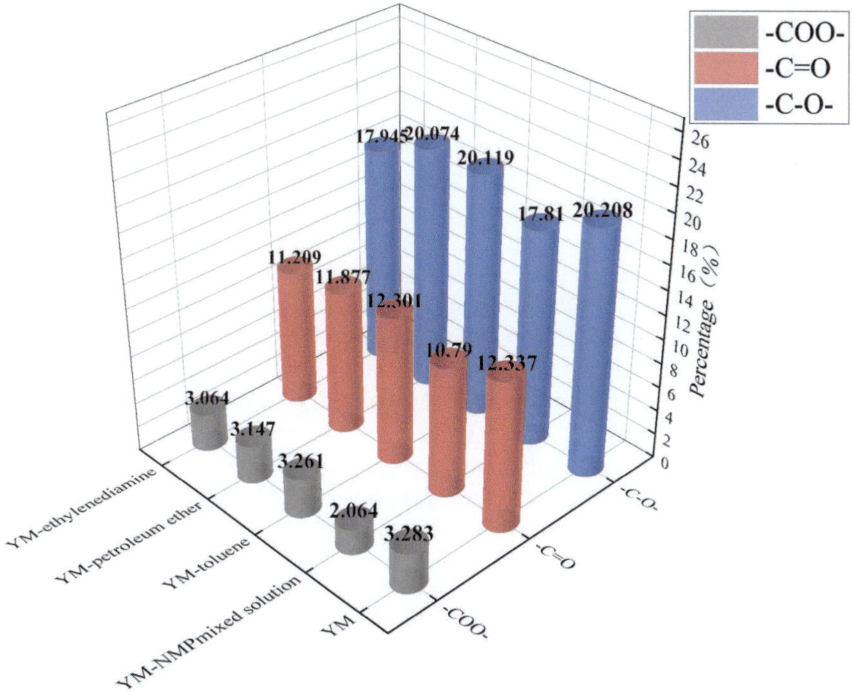

Fig. 4.6 Comparison of carbon and oxygen group content between YM and its four extracted coal samples

interaction, which results in the poor effect of petroleum ether on the molecular model of the side-chain of the carbonyl oxygen group extraction. The lowest percentage of carbon and oxygen groups in the coal sample was 31.045%, which was 3.61% lower than that of the original YM coal, due to the high electronegativity of the O and N atoms contained in the NMP molecule. This polarity enables NMP to have stronger interactions with the carbon and oxygen groups in the coal, forming two types of hydrogen bonds with stronger interactions, N–H–O and O–H–O. This inference is consistent with the results of the isopotential surface plot analysis of $y = 0.5$ for NMP on coal molecules with carbon and oxygen groups in the weak interaction analysis in Chap. 3, and the stronger interactions imply that NMP has a stronger selectivity for the carbon and oxygen groups. Therefore, NMP has better extraction ability for coal molecule model with carbonyl oxygen group side chain, which is shown as the lowest percentage of carbonyl oxygen group in NMP extracted coal samples in the infrared spectra analysis results.

(c) **Analysis of the percentage of aliphatic groups after YM extraction**

The groups with absorption peaks located between 2700 and 3000 cm^{-1} were mainly attributed to aliphatic groups. Figure 4.7 shows the results of peak fitting of the absorption peaks between 2700 and 3000 cm^{-1} for coal samples after YM extraction by different extractants. Integrating the peak splitting curves in Fig. 4.7, the contents of aliphatic groups in the four extracted coal samples were obtained, and then the contents of aliphatic groups between YM and its four extracted coal samples were plotted, and the results are shown in Fig. 4.8.

The content of aliphatic groups in the four extracted coal samples was obtained by integrating the peaks of the fitted peaks, and the comparison of the content of aliphatic groups between YM and its four extracted coal samples was plotted, as shown in Fig. 4.8. As shown in Fig. 4.8, the percentage of aliphatic groups in the four extracted coal samples after extraction with different extractants was ethylenediamine extracted coal sample (26.479%) > toluene extracted coal sample (26.176%) > NMP extracted coal sample (25.873%) > petroleum ether extracted coal sample (24.008%) in descending order. Among the four extractants, petroleum ether has the strongest extraction ability for aliphatic groups in YM molecules, and ethylenediamine has the weakest extraction ability.

The remaining aliphatic hydrocarbon groups of ethylenediamine-extracted coal samples are the highest among the four groups of samples, 26.479%, which is only 0.337% lower compared to the YM raw coal, this is due to the fact that the N atoms in

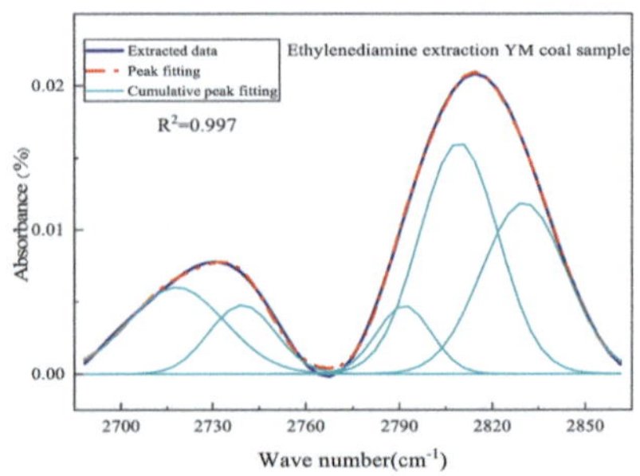

(a) Aliphatic group content of ethylenediamine extracted YM coal samples

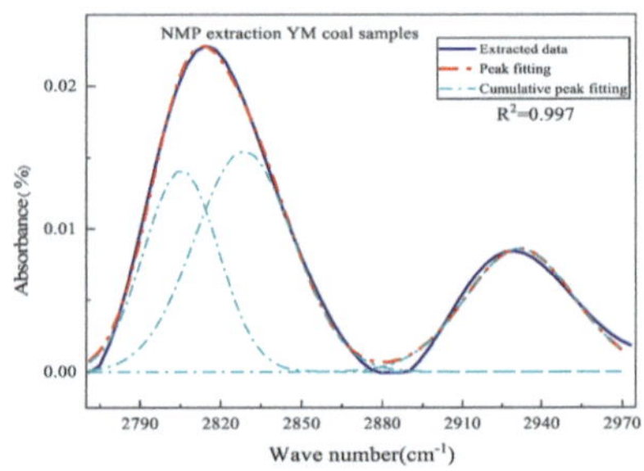

(b) Aliphatic group content of NMP extracted YM coal samples

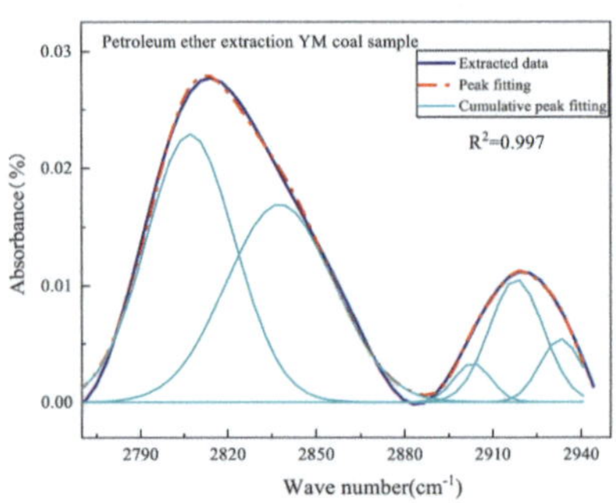

(c) Aliphatic group content of petroleum ether extracted YM coal samples

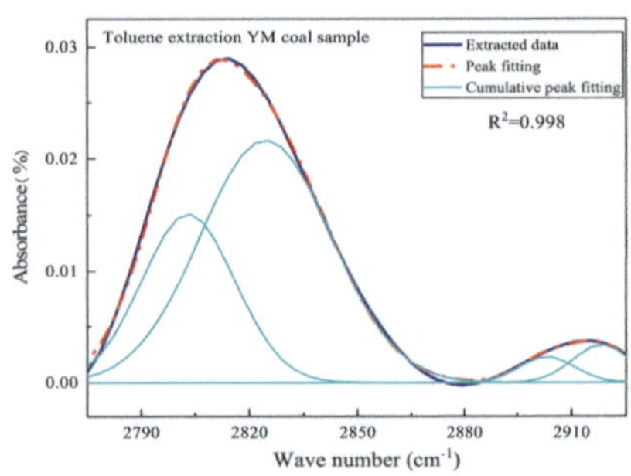

(d) Aliphatic group content of toluene extracted YM coal samples

Fig. 4.7 Fitting of aliphatic group peaks of four extracted coal samples

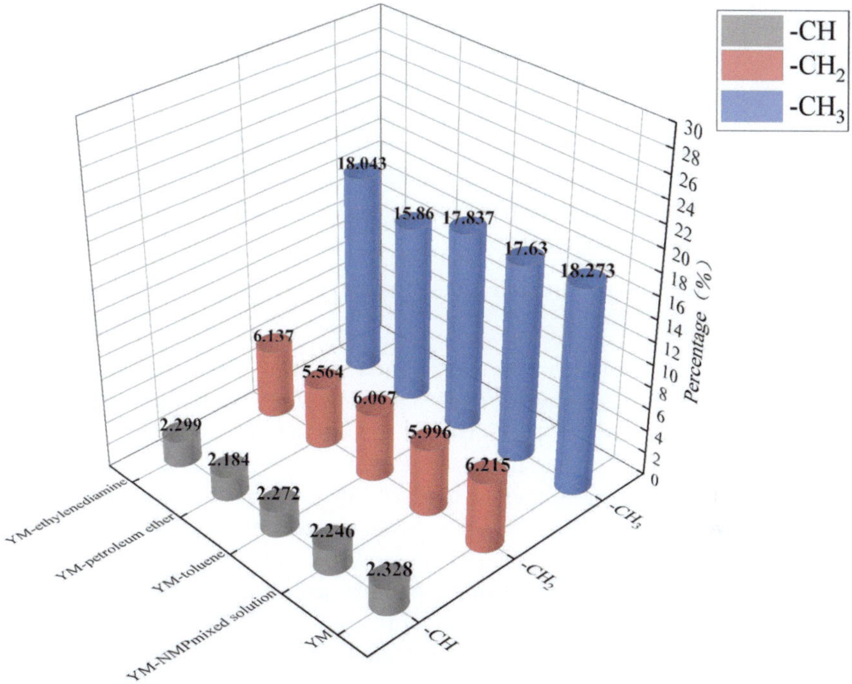

Fig. 4.8 Comparison of aliphatic group content between YM and its four extracted coal samples

ethylenediamine have lone pairs of electrons available for the formation of hydrogen bonding, but aliphatic hydrocarbons usually lack the atoms or groups available for the formation of hydrogen bonding, therefore, there does not exist any form of strong hydrogen bonding between ethylenediamine and aliphatic hydrocarbons, and this deduction can be verified by the molecular model interaction isosurface diagrams in the This inference can be verified by the isosurface plots of the interaction between ethylenediamine and the coal molecular model with aliphatic side chains in the weak interaction analysis in Chap. 3, so it is difficult for ethylenediamine to effectively extract the aliphatic hydrocarbon groups in coal through the attraction of hydrogen bonding; the percentage of aliphatic hydrocarbon groups in petroleum ether-extracted coal samples was the lowest, 24.008%, which is a decrease of 2.807% compared with that of the original coal of YM, which was attributed to the chromatography between the n-hexane and aliphatic groups in petroleum ether. This is due to the dispersion force between n-hexane and aliphatic groups in petroleum ether, which is a kind of weak interaction force between nonpolar molecules. Since n-hexane, the main component of petroleum ether, has a larger molecular mass, it carries more electrons, and a stronger dispersion force will be generated between it and aliphatic groups, and under the cumulative effect of a large number of intermolecular dispersion forces, petroleum ether and aliphatic groups have a certain degree of attraction, which promotes them to be separated from coal, and ultimately the aliphatic group is not separated. The petroleum ether has a certain attraction with the aliphatic groups under the cumulative effect of a large number of intermolecular dispersive forces, prompting them to be separated from the coal, which ultimately affects the aliphatic group extraction. By comparing the extraction effect of petroleum ether on the three aliphatic groups, it can be seen that $-CH_3$ has the largest decrease of 13.2%, which is due to its larger relative molecular mass compared with methylene and hypomethylene, and it is easier to be extracted because of its stronger interaction with n-hexane, the main component of petroleum ether.

(d) **Analysis of the proportion of aromatic hydrocarbon groups in YM extracted coal samples**

The absorption peaks located between 500 and 900 cm^{-1} were mainly attributed to aromatic hydrocarbon groups. Figure 4.9 shows the peak fitting results of the absorption peaks between 500 and 900 cm^{-1} for coal samples extracted by YM with different extractants. By integrating the peaks in Fig. 4.9, the content of aromatic hydrocarbon groups in the four extracted coal samples was obtained, and then the comparison of the aromatic hydrocarbon group content between YM and its four extracted coal samples was plotted, and the results are shown in Fig. 4.10.

The content of aromatic groups in the four extracted coal samples was obtained by integrating the split-peak curves of the fitted peaks, and then the comparison of the content of aromatic groups between YM and its four extracted coal samples was plotted, as shown in Fig. 4.10. As can be seen from the figure, after extraction with different extractants, the proportion of aromatic groups in the four extracted coal samples, from high to low, was ethylenediamine extracted coal sample (26.654%) > petroleum ether extracted coal sample (25.551%) > NMP extracted coal sample (24.117%) > toluene extracted coal sample (23.680%). Therefore, among the four extractants, toluene has the strongest extraction ability for the aromatic groups in YM molecules, and ethylenediamine has the weakest extraction ability. The remaining aromatic groups in the ethylenediamine-extracted coal samples were the highest among the four groups of samples, 26.654%, which was only 0.162% lower than that of the original coal of YM. This was due to the fact that the aromatic groups in YM have a high degree of stability, and the π–π stacking of the aryl rings with the aryl rings made the interaction between the molecular groups extremely stable and difficult to be broken, and the lack of sufficient selectivity of ethylenediamine for the aromatic compounds. In addition ethylenediamine lacks sufficient selectivity for aromatic compounds. Therefore, solvents with high selectivity are needed to effectively separate the target compounds. Toluene extracted coal samples have the lowest proportion of aromatic hydrocarbon groups, 23.680%, which is 3.136% lower than that of the original YM coal. This is due to the fact that toluene is a type of nonpolar organic solvent and is similar in structure to aromatic groups, which has a sound solubility to the aromatic groups in coal, so it can be used to extract such groups in coal molecules efficiently, and the interaction between toluene and the aromatic rings of the coal molecules can produce a π–π stack, which belongs to a strong van der Waals' solvent. This interaction is a strong van der Waals interaction, and the

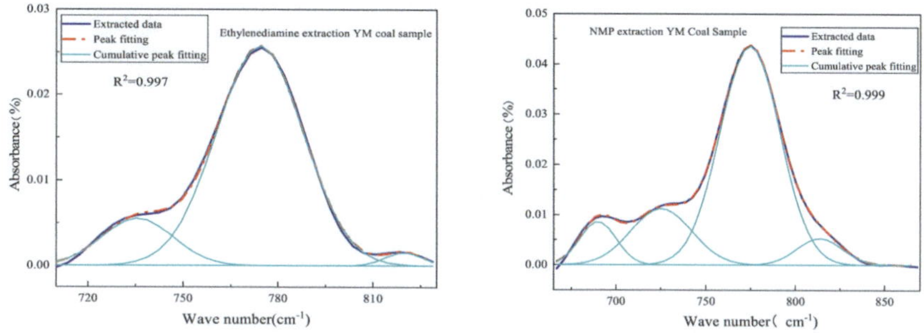

(a) Content of aromatic hydrocarbon groups in ethylenediamine extracted YM coal samples

(b) Aromatic hydrocarbon group content of NMP extracted YM coal samples

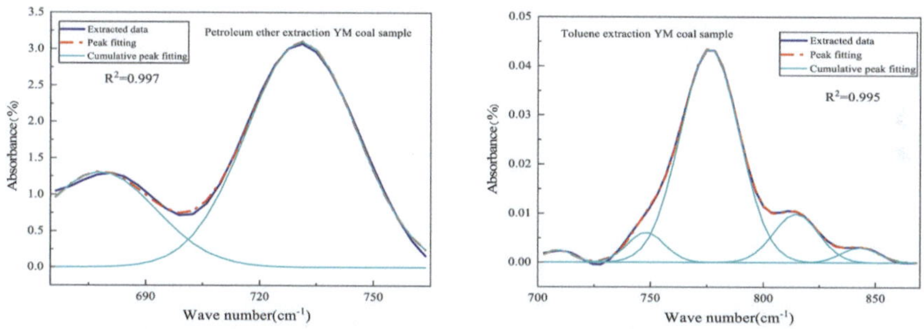

(c) Aromatic hydrocarbon group content of petroleum ether extracted YM coal samples

(d) Aromatic hydrocarbon group content of toluene extracted YM coal samples

Fig. 4.9 Aromatic hydrocarbon group peak fitting for four extracted coal samples

dispersive force generated by it makes the fragments of coal molecules containing aromatic groups more inclined to be assigned to toluene molecules.

4.2.2 *Infrared Spectrum Analysis Before and After WYM Extraction*

Figure 4.11 shows the infrared absorption spectra of coal samples extracted by WYM with four different extractants: Ethylenediamine, NMP, petroleum ether and toluene. In order to further clarify the content changes of different kinds of functional groups in the extracted coal samples after WYM extraction with different extractants, this paper classifies the spectrum diagram in Fig. 4.11 into four categories: Hydroxyl (–OH) groups, carbo-oxygen groups, aliphatic groups and aromatic groups according to the peak analysis results in Chap. 2. Therefore, in this paper, according to Fig. 4.11, the infrared absorption spectra of raffinizes coal samples were fitted by wave band and peak.

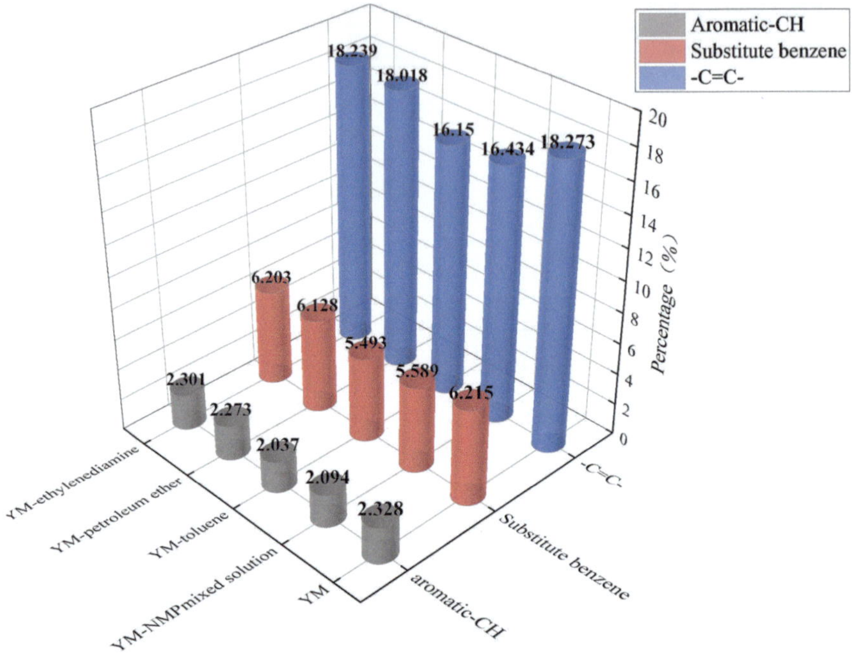

Fig. 4.10 Comparison of aromatic group content between YM and its four extracted coal samples

To further clarify the content changes of different groups in the extracted coal samples after WYM extraction by different extractants, the infrared spectrum was divided into four parts according to the infrared spectrum analysis in Chap. 2, namely hydroxyl (–OH) groups, carboxyl groups, aliphatic groups, and aromatic groups. The peak fitting and analysis results are as follows:

(a) **Analysis of hydroxyl (–OH) group proportion of WYM raffinate coal samples**

Groups with absorption peaks between 3000 and 3700 cm^{-1} are mainly attributed to –OH. Figure 4.12 shows the peak-to-peak fitting results of the absorption peaks between 3000 and 3700 cm^{-1} of coal samples extracted by WYM with different extractants. The peak curve in Fig. 4.12 was integrated to obtain the content of –OH groups in the four raffinate coal samples, and then the comparison of the content of –OH functional groups between WYM and the four raffinate coal samples was drawn, as shown in Fig. 4.13.

By integrating and calculating the sub-peak curves of the fitted peaks, the peak areas of different peaks were obtained respectively, and the peak areas were corresponding to the content of the peak groups, so as to draw a comparison diagram of hydroxyl –OH group content between WYM and four types of raffinative coal samples, as shown in Fig. 4.13. It can be seen from the figure that after extraction with different extractant, the proportion of OH groups in four kinds of raffinate

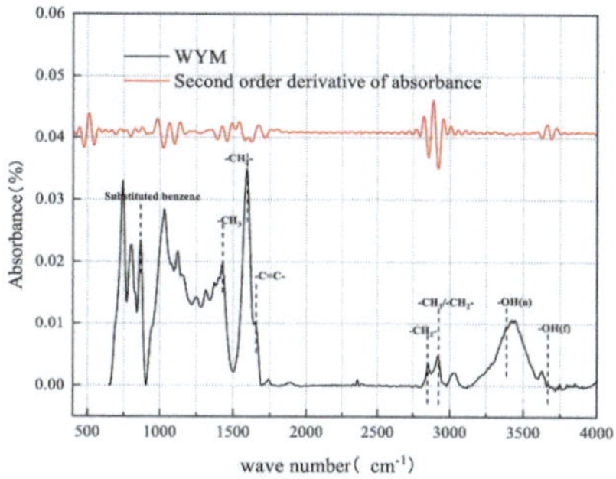

(a)WYM infrared spectrum

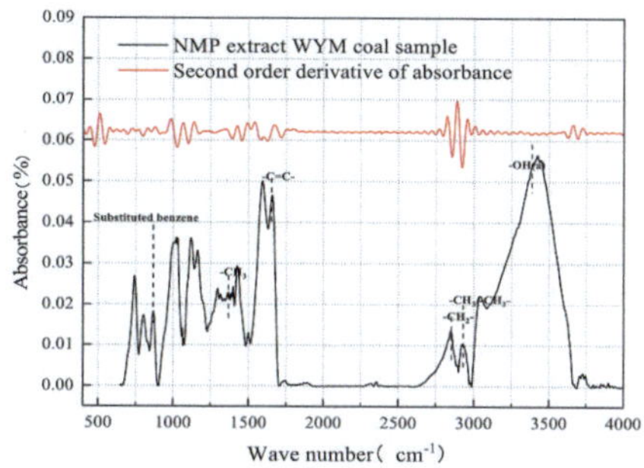

(b)Infrared spectrum of NMP extractive WYM coal sample

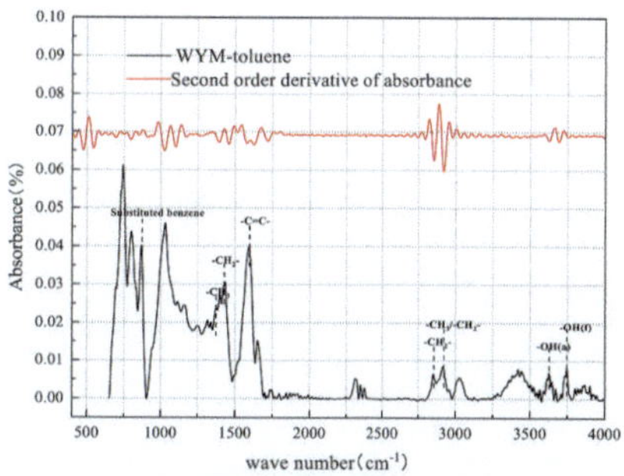

(c)Infrared spectrum of WYM coal samples extracted by toluene

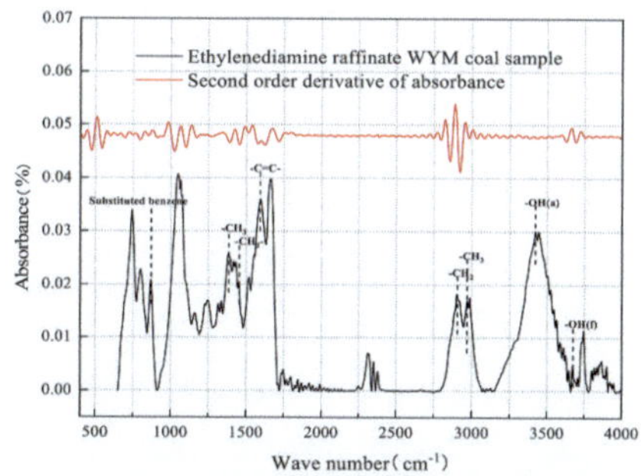

(d)Infrared spectrum of ethylenediamine raffinate WYM coal sample

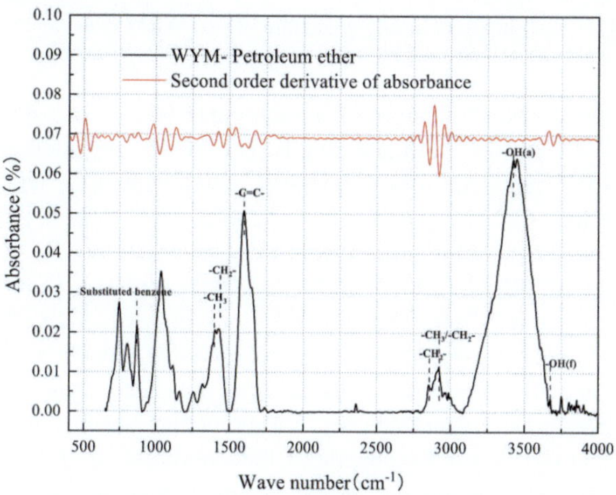

(e)Infrared spectra of WYM coal samples extracted from petroleum ether

Fig. 4.11 Infrared spectra of WYM and its raffinative coal samples

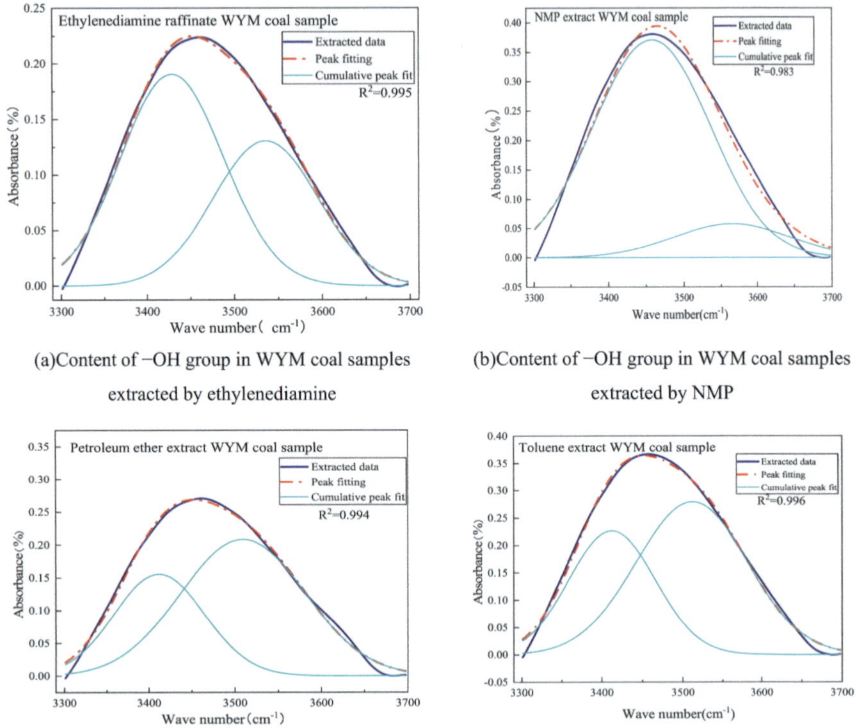

(a)Content of –OH group in WYM coal samples extracted by ethylenediamine

(b)Content of –OH group in WYM coal samples extracted by NMP

(c)Content of –OH group in WYM coal samples extracted from petroleum ether

(d)Content of –OH group in WYM coal samples extracted by toluene

Fig. 4.12 Hydroxyl (–OH) group peak fitting diagram of four WYM raffinate coal samples

samples from high to low is toluene raffinate sample (6.98%) > petroleum ether raffinate sample (6.328%) > NMP raffinate sample (5.731%) > ethylenediamine raffinate sample (5.191%). Among the four extractants, ethylenediamine has the strongest ability to extract the –OH group in WYM molecule, while toluene has the weakest ability.

The proportion of –OH in toluene raffinated coal samples is the highest among the four groups, which is 6.98%. The reason for the poor extraction effect of toluene on –OH in WYM is that toluene is a non-polar solvent, does not form hydrogen bond with –OH in coal molecular groups, and can only form a weak van der Waals interaction. Therefore, its extraction effect on molecular groups containing –OH is not as good as that of other polar molecules. Second, although NMP increases the polarity of the solvent system to a certain extent, it may not be enough to fully interact with the –OH in WYM. The proportion of –OH group in ethylenediamine raster coal sample is the lowest in the five groups, which is 5.191%. This is because ethylenediamine, as a typical fat diamine, has sclient polarity characteristics. The N atom on the $–NH_2$ has strong electronegativity, which can form H-bond receptor,

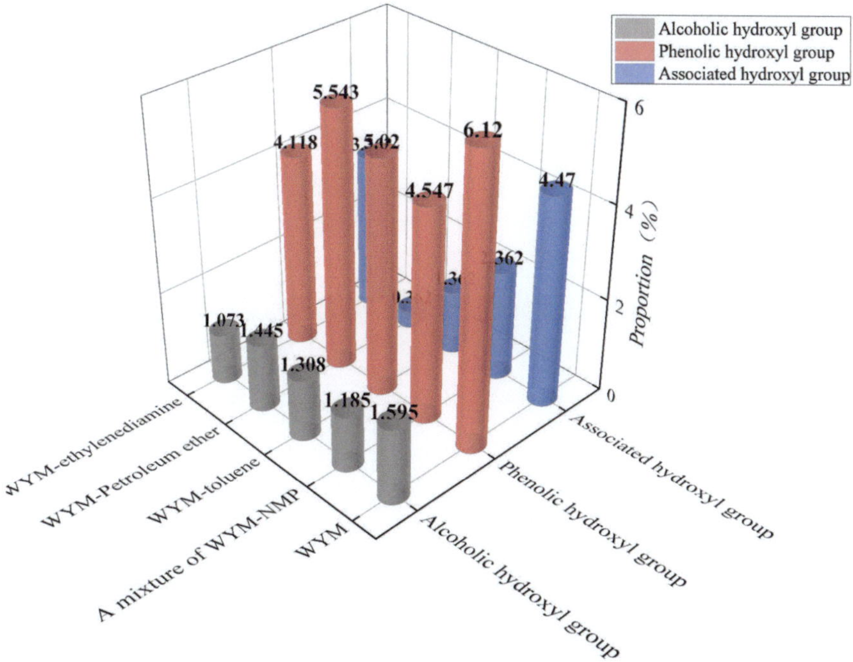

Fig. 4.13 Comparison of –OH group content between WYM and four raffinate coal samples

and N–H–O hydrogen bond is formed between the O atom in –OH. This allows it to interact effectively with the –OH group, which can be verified by the analysis of the isopotential diagram of the interaction of ethylenediamine with the –OH side chain in the coal molecule in Chap. 3. The amino group in ethylenediamine molecule can form hydrogen bond with –OH, which enhances the interaction force between them and is conducive to the extraction of –OH from WYM.

(b) **Analysis of carbon and oxygen groups in WYM raffinate coal samples**

The groups with absorption peaks between 1500 and 1750 cm^{-1} are mainly attributed to the carbon–oxygen groups. Figure 4.14 shows the peak-to-peak fitting results of the absorption peaks between 1500 and 1750 cm^{-1} of coal samples extracted by WYM with different extractants. The peak curve in Fig. 4.14 was integrated to obtain the content of carbon and oxygen groups in four kinds of raffinate coal samples, and then the comparison diagram of the content of carbon and oxygen groups between WYM and four types of raffinate coal samples was drawn, and the results are shown in Fig. 4.15.

The content of carbon and oxygen groups in four types of raffinate coal samples was obtained by integrating the sub-peak curves of the fitted peaks, and the comparison diagram of the content of carbon and oxygen groups between WYM and four kinds of raffinate coal samples was drawn, as shown in Fig. 4.15. It can be seen from the figure that after extraction with different extracents, the proportion of carbon

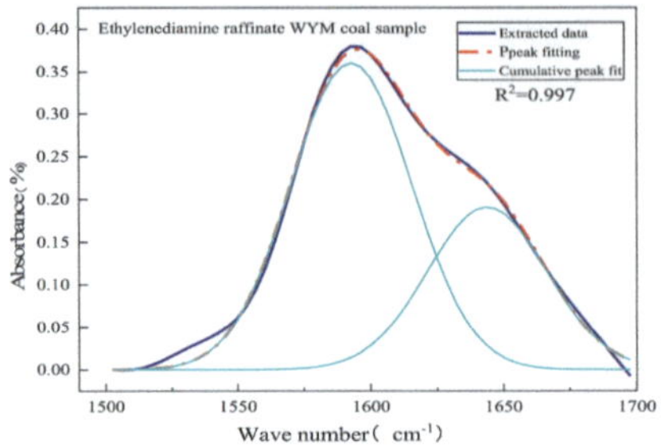

(a) The content of carbo-oxygen groups in WYM coal samples

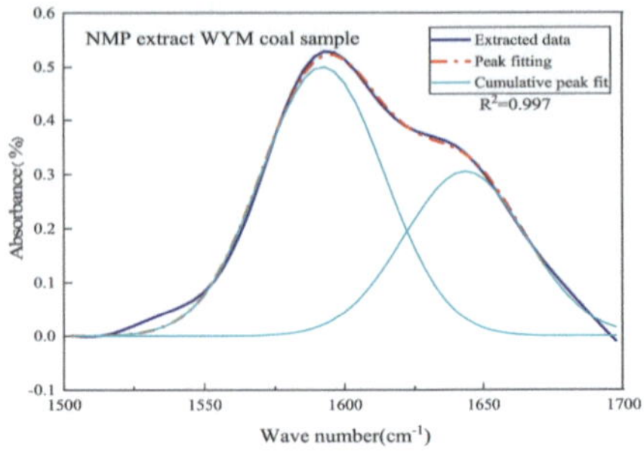

(b)Content of carbon oxygen groups in WYM coal samples extracted by NMP

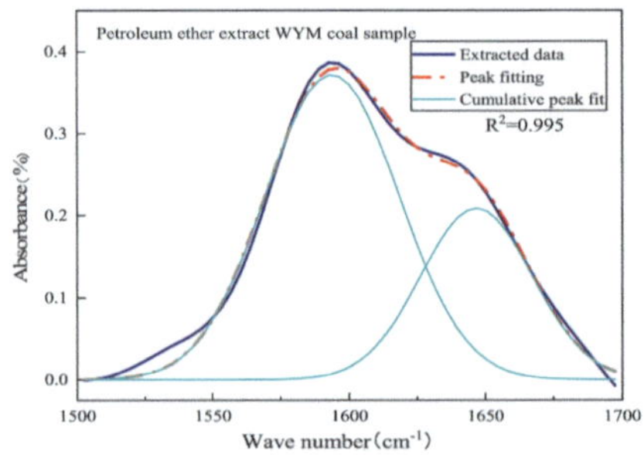

(c)Content of carbon and oxygen groups in WYM coal samples extracted from petroleum ether

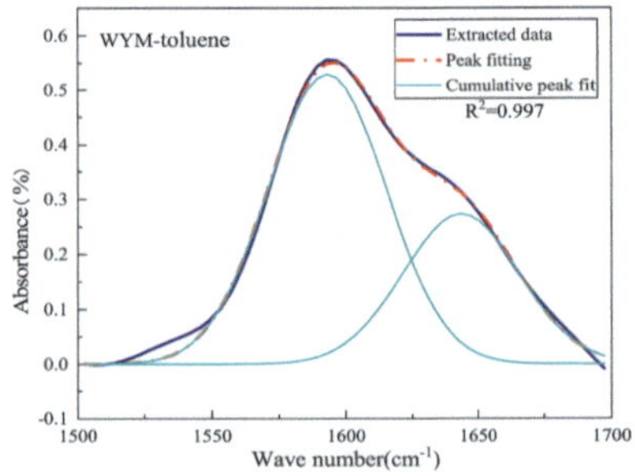

(d)Content of carbon and oxygen groups in WYM coal samples extracted by toluene

Fig. 4.14 Fitting diagram of carbon and oxygen group peaks of four WYM raffinative coal samples

oxygen groups in four types of raffinate samples from high to low is toluene raffinate sample (24.410%) > petroleum ether raffinate sample (24.065%) > ethylene diamine raffinate sample (22.232%) > NMP raffinate sample (21.783%). Therefore, among the four extractants, NMP has the strongest extraction ability for carbon and oxygen groups in WYM, while toluene has the weakest extraction ability.

The ratio of carbon and oxygen groups in toluene raffinated coal samples was the highest among the five groups, which was 24.410%, and only decreased by 0.14% compared with raw coal. The main reason for the poor effect of toluene extraction of carbon and oxygen groups in WYM was that toluene, as a non-polar molecule, did not contain any strong electronegative atoms and could not form H receptors for O atoms in carbon and oxygen groups. There is only a weak interaction force between the two molecules, the van der Waals force, and the high carbon content and tight structure of WYM itself cause difficulty in solvent penetration, which further increases the difficulty of extraction. The ratio of carbon and oxygen groups in NMP raffinate coal samples is the lowest among the five groups, which is 21.783%. This is because NMP is a polar solvent, and the N atom on its ring will form an H bond receptor due to its electronegativity, which can form N–H–O hydrogen bond with the O atom in the carbon and oxygen groups. This inference is consistent with the

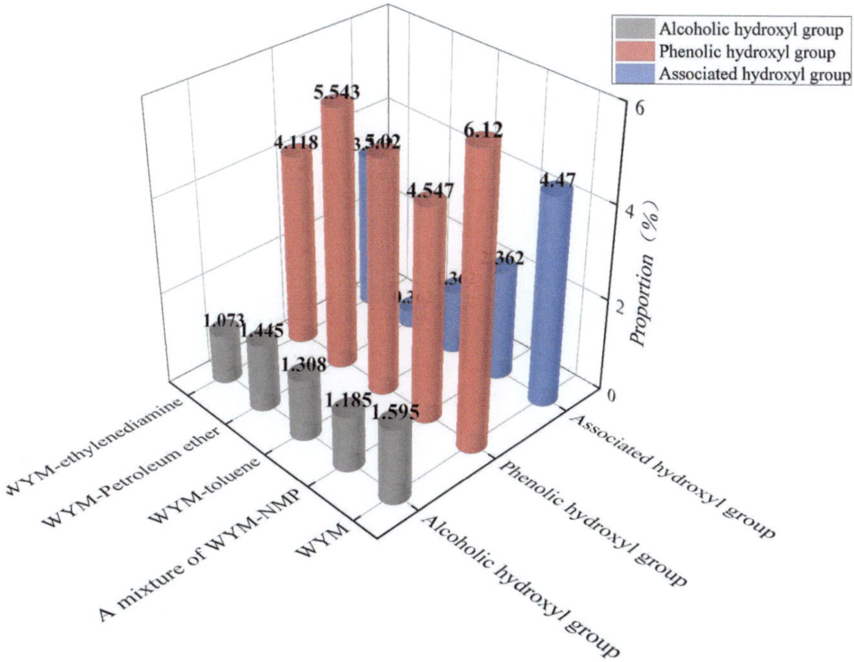

Fig. 4.15 Comparison of carbon and oxygen group content between WYM and four raffinate coal samples

isopotential diagram of the interaction between NMP and the coal molecular model with carbon and oxygen groups in the side chain in the weak interaction analysis in Chap. 3. Moreover, NMP has a strong dissolution ability, which enables NMP to effectively interact with the molecular model with carbon and oxygen groups in WYM, and thus realize the extraction of such groups.

(c) Analysis of aliphatic group proportion of WYM raffinate coal sample

Groups with absorption peaks between 2700 and 3000 cm^{-1} are mainly aliphatic groups. Figure 4.16 shows the peak-to-peak fitting results of the absorption peaks between 2700 and 3000 cm^{-1} of coal samples extracted by WYM with different extractants. The peak curve in Fig. 4.16 was integrated to obtain the aliphatic group content in four types of raffinate coal samples, and then the comparison diagram of the aliphatic group content between WYM and four types of raffinate coal samples was drawn, and the results were shown in Fig. 4.17.

The content of carbo-oxygen groups in the four raffinate coal samples was obtained by integrating the sub-peak curves of the fitted peaks, so as to draw a comparison diagram of aliphatic group content between WYM and the four raffinate coal samples, as shown in Fig. 4.17. It can be seen from the figure that the aliphatic groups of the four types of raffinate samples after extraction with different extracents are from high to low as ethyleneamine raffinate sample (38.145%) > NMP raffinate sample

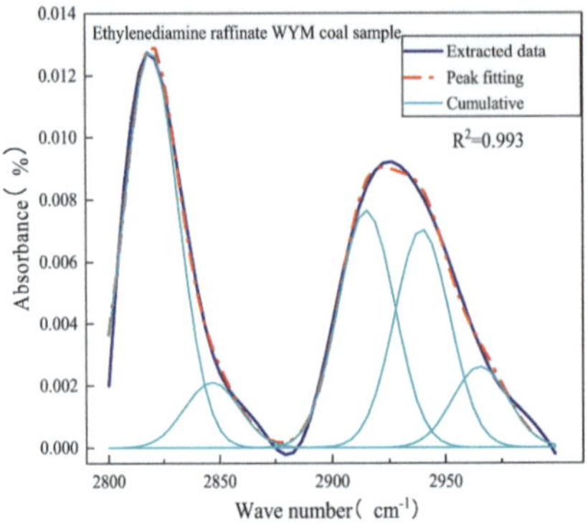

(a)Proportion of aliphatic groups in the WYM coal sample

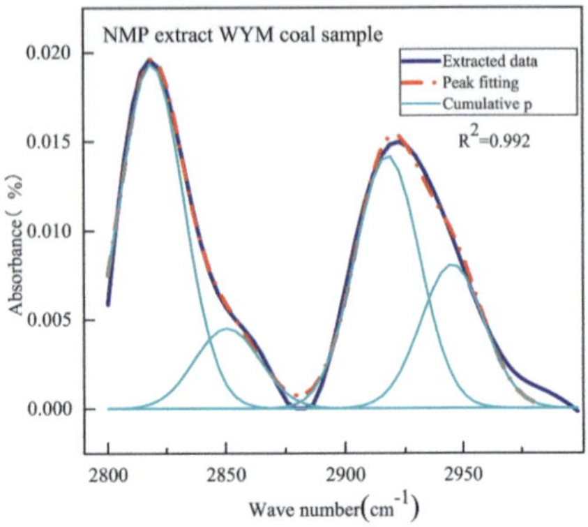

(b)Proportion of aliphatic groups in NMP raffinate WYM coal samples

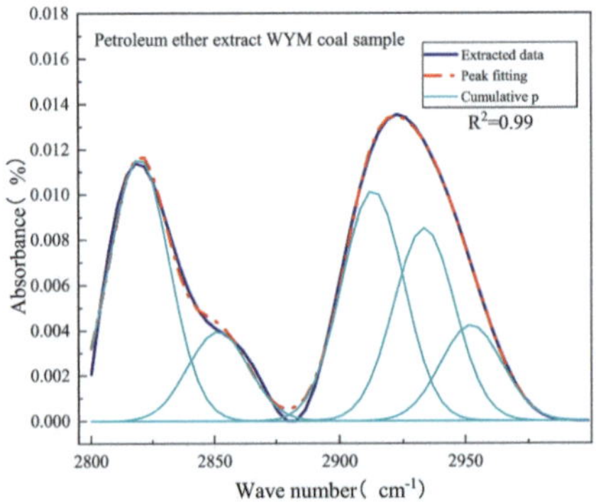

(c)Proportion of aliphatic groups in coal samples extracted by petroleum ether

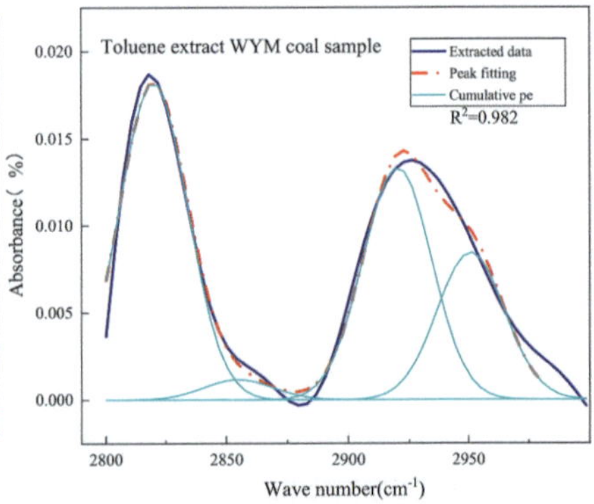

(d)Proportion of aliphatic groups in WYM coal samples extracted by toluene

Fig. 4.16 Aliphatic group peak fitting diagram of four WYM raffinite samples

(37.174%) > toluene raffinate sample (35.715%) > petroleum ether raffinate sample (34.295%). Among the four extractants, petroleum ether has the strongest ability to extract aliphatic groups from WYM molecules, while ethylenediamine has the weakest ability.

The proportion of aliphatic hydrocarbons in ethylenediamine raffinate coal samples is the highest among the five groups, which is 38.145%. The main reason for the poor extraction effect of aliphatic hydrocarbons in WYM is that compared with other carbo-oxygen groups, aliphatic groups usually lack hydrogen bond donors or receptors. Therefore, there is no hydrogen bond between ethylenediamine and aliphatic groups, and the influence of strong interaction force is lacking. It is difficult for ethylenediamine to extract the coal molecular groups containing aliphatic

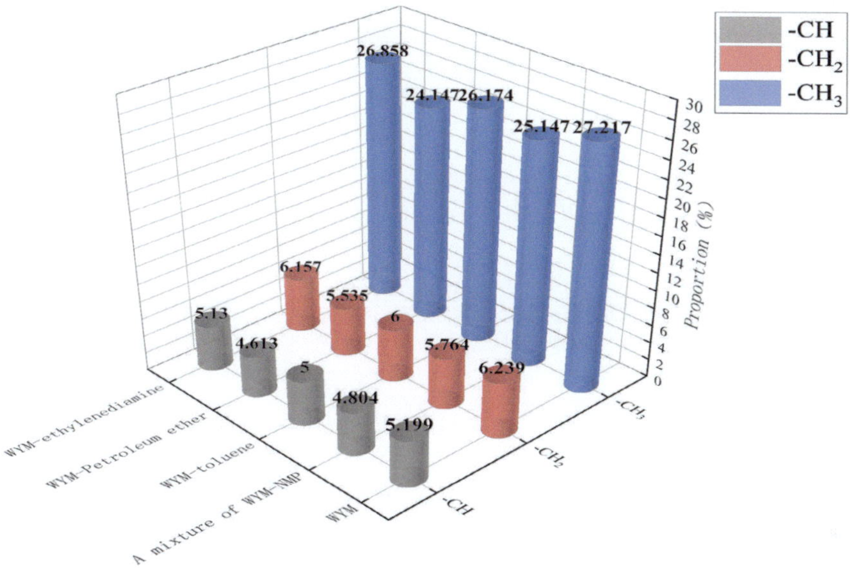

Fig. 4.17 Comparison of aliphatic group content between WYM and its four raffinate coal samples

side chains. The proportion of aliphatic groups in the petroleum ether raffinate coal samples is the lowest in the five groups, which is 34.295%, because the main component of petroleum ether, n-ethane, is a non-polar molecule, and aliphatic groups are usually non-polar structures composed of carbon and hydrogen. This similarity makes n-hexane have a superior ability to dissolve aliphatic groups, and n-hexane has a relatively high molecular weight compared to the four extracants, and the instantaneous dipole moment generated by its electron cloud is stronger, thus forming a strong dispersion force, so that n-hexane can extract aliphatic groups from coal, and the proportion of aliphatic groups in raffinate coal samples is lowest.

(d) **Analysis of the proportion of aromatic groups in WYM raffinaceous coal samples**

Groups with absorption peaks between 500 and 900 cm^{-1} mainly belong to aromatic hydrocarbon groups. Figure 4.18 shows the peak fitting results of absorption peaks between 500 and 900 cm^{-1} of coal samples extracted by WYM with different extractants. By integrating the peak curves in Fig. 4.18, the contents of aromatic hydrocarbon groups in four raffinate coal samples were obtained, and then the comparison of the content of aromatic hydrocarbon groups between WYM and four raffinate coal samples was drawn, as shown in Fig. 4.19.

The content of aromatic groups in the four raffinative coal samples was obtained by integrating the sub-peak curves of the fitted peaks, so as to draw a comparison diagram of the content of aromatic groups between WYM and the four raffinative coal samples, as shown in Fig. 4.19. It can be seen from the figure that after extraction with different extractants, the proportion of aromatic groups in four kinds of raffinate samples from

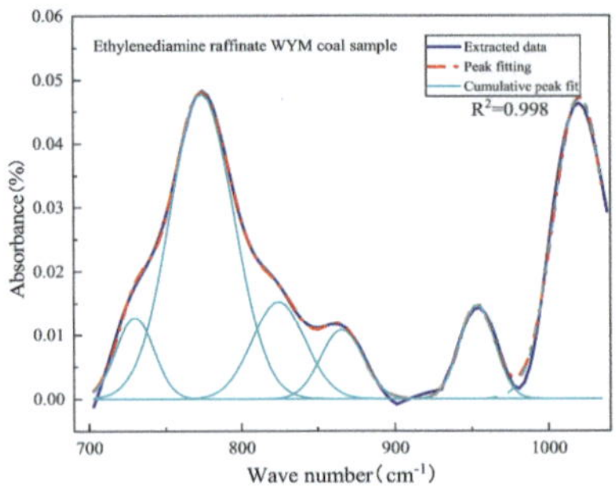

(a)Aromatic group content of ethylenediamine raffinate WYM coal

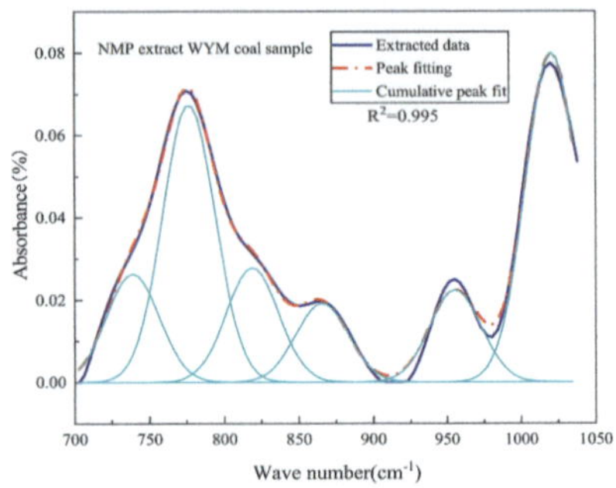

(b)Content of aromatic groups in WYM coal samples extracted by NMP

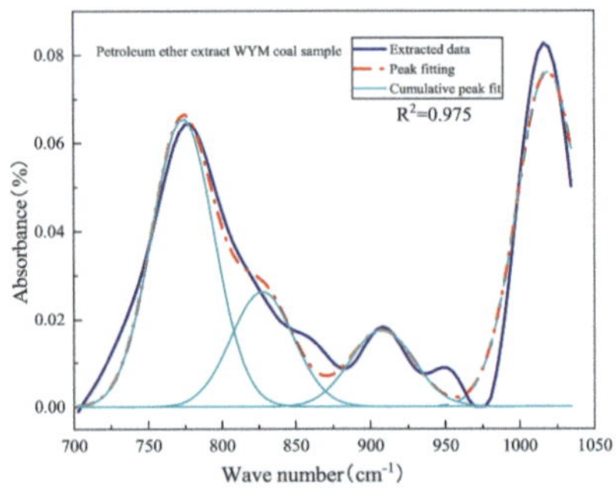

(c)Content of aromatic groups in WYM coal samples extracted from petroleum ether

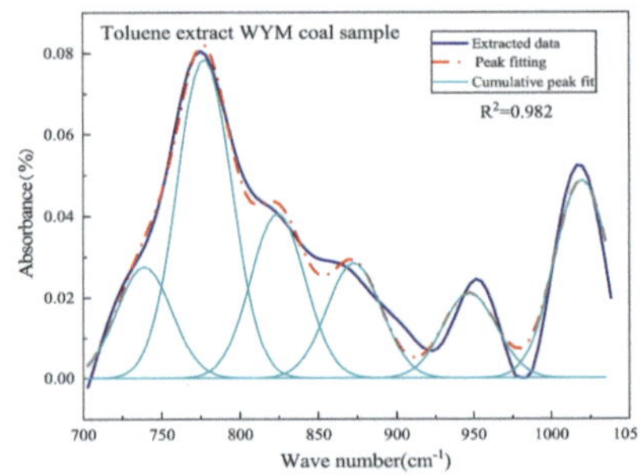

(d)Content of aromatic groups in WYM coal samples extracted by toluene

Fig. 4.18 Aromatic group peak fitting diagram of four WYM raffinite samples

high to low is ethylenediamine raffinate sample (38.426%) > NMP raffinate sample (38.323%) > petroleum ether raffinate sample (36.117%) > toluene raffinate sample (35.010%). Therefore, among the four extractants, toluene has the strongest ability to extract aromatic groups from WYM molecules, while ethylenediamine has the weakest ability.

The proportion of aromatic hydrocarbons in ethylenediamine raffinate coal sample is the highest among the five groups, which is 38.426%, which is only 0.229% lower than that of WYM raw coal. The main reason for the poor effect of ethylenediamine extraction of aromatic hydrocarbons in WYM is that the aromatic groups are usually composed of aromatic rings, and the interaction with ethylenediamine is relatively weak. As a result, coal molecular groups containing aromatic side chains are usually not selected for allocation to ethylenediamine organic solvents. Therefore, the selectivity of ethylenediamine to molecular groups containing aromatic groups is poor, and it is difficult to extract it from coal effectively. The proportion of aromatic groups in toluene solvent extract coal samples is the lowest in the five groups, 35.010%, which is 3.64% lower than that of WYM raw coal. This is because toluene and coal are both

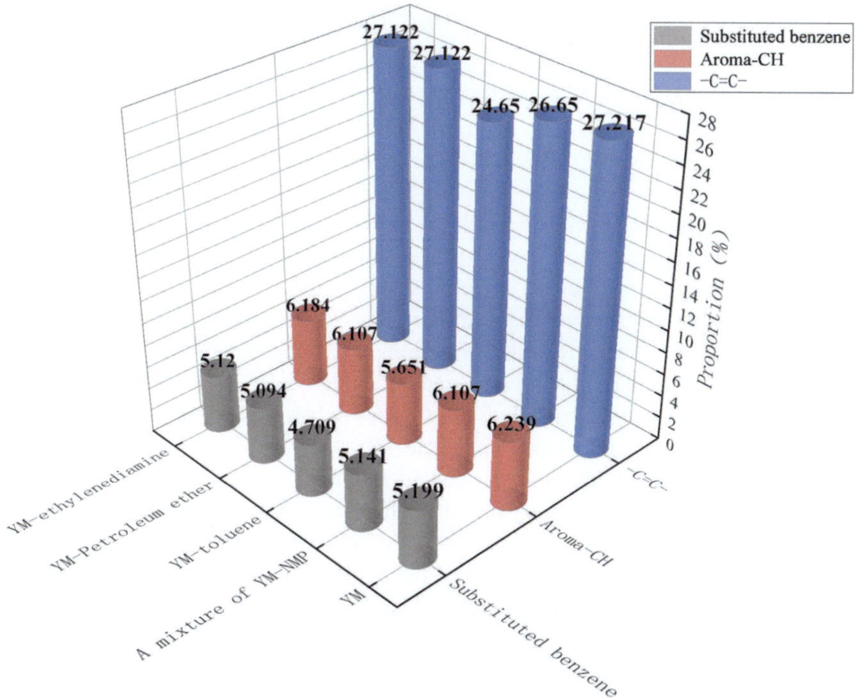

Fig. 4.19 Comparison of aromatic group content between WYM and four raffinaceous coal samples

polar molecules, and the aromatic ring structure shared by toluene and coal enables π–π accumulation, forming a strong van der Waals attraction. Thus, the transfer and separation of coal molecular groups containing aromatic side chains can be realizes, and the proportion of aromatic groups in raffinate coal samples can be minimizes.

4.3 Characterization of Free Radical Changes in Coal Samples Before and After Extraction

In the process of oxidative spontaneous combustion of coal, radicals are important factors affecting the intensity of the reaction, and in the radical chain reaction, the concentration and changes of radicals in the coal samples determine whether the samples have the ability to spontaneous combustion or not, therefore, this paper measured the radicals of the coal samples before and after extraction, and investigated the changes of radicals in the coal samples before and after extraction by different extractants.

4.3.1 Experimental Apparatus and Methods

In this paper, the Bruker EMXplus-9.5/12 EPR (Schmallegger and Gescheidt 2018) from Germany was used to test the variation of radical concentration in raw and extracted coal samples from YM and WYM, and the experimental setup is shown in Fig. 4.20.

In the experiment, 10 mg of solid products produced by different extractants were weighed each time and loaded into a brand new 5 mm Norell NMR tube, which was subjected to radicals determination by using Bruker EMXplus-9.5/12 EPR as shown in the figure: The peak widths, g-values, quadratic integral areas were recorded, and the radical concentrations were calculated. Among them, the data quantitative calculation will be carried out using Bruker's own Xenon software, which can directly obtain the absolute quantitative radical concentration. The specific experimental parameters are shown in Table 4.3.

In addition, this experiment also considered the position of the sample tube inserted into the resonance cavity and the influence of different batches of sample tubes on the experimental results (Fu et al. 2024), the same batch of sample tubes was used in the experimental process and the depth of the sample tube inserted into the resonance cavity was 6 cm for each measurement.

Fig. 4.20 BrukerEMXplus-9.5/12EPR, Germany

Table 4.3 Experimental parameters for EPR testing

Parameter	Numerical
Test temperature	298 K
Microwave frequency	9.48 GHz
Microwave power	0.2 W
Microwave attenuation	30 dB
Center magnetic field	3464.6 G
Scanning width	200 G
Scanning time	30 s
Time constant	10.24 ms
Modulation frequency	100 kHz
Modulation amplitude	1 G

4.3.2 *Comparative Analysis of Radical Characteristics of YM Raw Coal and WYM Raw Coal*

Usually EPR spectra are composed of g-value and intensity, and the transverse g-value is calculated with reference to (Eq. 4.2):

$$g = hv/\beta H \tag{4.2}$$

In the formula: $h = 4.135 \times 10^{-15}$ eV is Planck's constant; $v = 9.48$ GHz is the microwave frequency of the x-band spectrometer; $\beta = 5.788 \times 10^{-5}\mathrm{evT}^{-1}$ is the electron Bohr flux; and H is the applied magnetic field.

The EPR experimental data were plotted to obtain the EPR spectrograms of YM and WYM, as shown in Fig. 4.21.

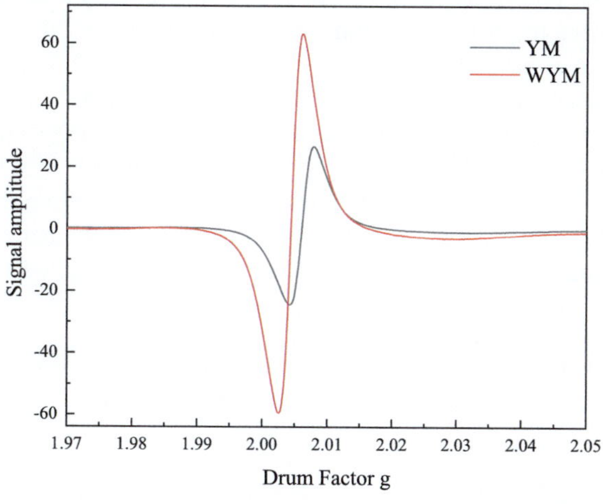

Fig. 4.21 EPR spectra of YM and WYM coal samples

The analysis of an EPR spectrogram usually consists of three aspects:

(1) Radical concentration: The number of free electrons per gram of sample;
(2) Drum factor (g-factor): It determines the position of the spectral lines in the EPR spectra and is a direct factor reflecting the degree of spin-orbit interference;
(3) Line width (ΔH), which is determined by the lattice relaxation time.

The individual parameters of the EPR spectra of YM and WYM are shown in Table 4.4.

From Table 4.4, it can be seen that the radical Delang factor g-values of YM and WYM are 2.00268 and 2.00259, respectively, which are greater than the g-factor value of free electrons, 2.0023. This is due to two factors, one is that the radical g-values of aromatics or chain hydrocarbons, whose electrons are in π-orbitals, are 2.0035 and 2.0046, respectively, which are the main constituents of coal molecules, and the other is that the radical g-factor of coal is 2.0023, which is the main constituent of coal molecules. Therefore the radical g-factor of coal must be greater than the g-factor value of free electrons, which is above 2.0023. And according to the concentration of aromatic and chain hydrocarbon radicals, it can be seen that the higher the content of aromatic hydrocarbons in the coal, the smaller the radical drones factor g of the coal; the higher the content of chain hydrocarbons, the larger the value of the radical drones factor g. Since the degree of metamorphism of WYM is higher than that of YM, the value of the g factor is even smaller. On the other hand, the g factor of heteroatoms contained in the coal are all larger than the g factor value of free electrons (the g value of N radicals is 2.0031, and the g value of S radicals is 2.004), so the coal with a lower degree of metamorphism has a larger g factor due to the presence of more heteroatoms, and according to the results of the industrial analysis in Chap. 2, the YM selected in this paper is a high-sulfur coal with a higher sulphur content, and therefore the g factor value of YM is greater, further verifying the accuracy of the EPR test.

The linewidth is inversely proportional to the relaxation time of the electrons, i.e., the time it takes for the electrons in a substance to return to the equilibrium state from the nonequilibrium state, and as shown in Table 4.4, the linewidths of YM are larger than those of WYM, which are 7.17 and 5.91, respectively. The relaxation time can characterizes the difficulty of the reaction to a certain extent because the shorter the relaxation time is, it means that the system responds to the external perturbation faster, and it can be restored to the equilibrium state more quickly to equilibrium. In this case, it may be easier for the system to jump from the ground state or other low-energy state to the excited state because the barrier of energy is lower and the activation energy is smaller, so the energy required for the reaction is lower, and

Table 4.4 Parameters of EPR spectrograms of YM and WYM coal samples

Sample	Drum factor g	Line width ΔH (G)	Radical concentration C (10^{17}/g)
YM	2.00268	7.17	11.29
WYM	2.00259	5.91	72.53

therefore the reaction is more likely to occur. However, the ease of spontaneous combustion of coals with different degrees of metamorphism cannot be analyzes simply by the length of the linewidth, because the length of the linewidth is affected by both spin–lattice relaxation and spin relaxation, and the relaxation time of WYM is longer than that of YM, mainly due to the fact that the aromatic condensation of WYM is more advanced, which prolongs the relaxation time of the electron spin–lattice interactions, whereas the relaxation time of the non-spin relaxation is longer, so the WYM linewidth is longer than YM, so the WYM linewidth is longer. longer, so the WYM linewidth is longer than that of YM, yet it is less prone to spontaneous combustion. However, when the relaxation times of the same coal are comparatively analyzes, the conclusion that the shorter the time, the more readily the coal is oxidizs can be used to explain part of the phenomenon.

From Table 4.4, it can be seen that the radical concentration of WYM is higher than that of YM. Based on previous studies, it is known that a large number of alkanes are detached from the side chain during the deterioration of coal, and the alkane fragments and the originally connected structural molecules each have an unpaired electron, i.e., a free electron, and the concentration of radicals rises if the increase in the unpaired electrons is less than the decrease in the number of their occurrence of pairing, so the increase in aromatic hydrocarbon moieties makes the WYM has a greater concentration of radicals than YM.

4.3.3 Characteristics of Free Radical Changes Before and After YM Extraction

The EPR experimental curves of coal samples before and after YM extraction were plotted according to the experimental data, and the changes of g-factor, line width and radical concentration were investigated separately. As shown in Fig. 4.22, the specific EPR parameters are shown in Table 4.5.

As can be seen from Fig. 4.22, the radical concentrations, species and properties in the extracted coal samples of YM changed to different degrees after extraction with ethylenediamine, NMP, petroleum ether and toluene, respectively. To further analyzes the reasons for the changes, the calculated Drum factor g, line width ΔH and radical concentration C are summarizes in Table 4.5.

As shown in Table 4.5, the YM delta factor (g) is larger than that of ethylenediamine and NMP extracted coal samples, but smaller than that of toluene and petroleum ether extracted coal samples, which indicates that the contents of –OH and carbon–oxygen radicals in extracted coal samples are toluene extracted YM coal samples, petroleum ether extracted YM coal samples, NMP extracted YM coal samples, and ethylenediamine extracted YM coal samples, in descending order of their contents. In contrast, the contents of aromatic and aliphatic radicals in the extracted coal samples were ethylenediamine extracted, NMP extracted, petroleum ether extracted and toluene extracted in descending order. This conclusion is mainly

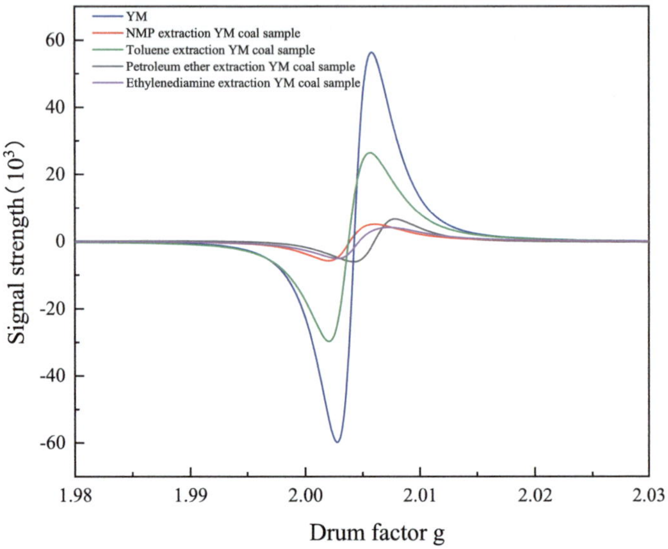

Fig. 4.22 EPR experimental curves of YM raw coal samples and extracted coal samples

Table 4.5 Parameters of EPR spectra of YM's extracted coal samples

Sample	Drum factor (g)	Line width ΔH (G)	Radical concentration C (10^{17}/g)
YM	2.00268	7.17	11.29
Ethylenediamine extraction YM coal sample	2.00251	5.38	4.34
NMP extract YM coal sample	2.00263	6.33	5.13
Petroleum ether extract YM coal sample	2.00273	7.49	6.21
Toluene extract YM coal sample	2.00286	7.61	9.66

due to the fact that toluene and petroleum ether can form van der Waals forces with aliphatic and aromatic groups in the coal molecule, which can partially detach the above two groups from the coal macromolecular system, resulting in an increase in the content of O atoms in the extracted coal samples, and the Drum factor (g) of O atoms is larger than that of C atoms, so that the Drum factor (g) of toluene and petroleum ether extracted YM samples is larger than that of YM. Meanwhile and the Drum factor (g) of toluene and petroleum ether extracted YM samples are larger than that of YM. In contrast, NMP and ethylenediamine will strip part of –OH and carbon–oxygen groups in coal molecules, which leads to the increase of aliphatic and aromatic ring content in the extracted coal samples, and according to the results of the previous studies, it can be seen that the free radicals g of aromatic hydrocarbons or chain hydrocarbons with electrons in π-orbitals are 2.0035 and 2.0046, respectively, which are both less than the YM delta factor, and thus the delta factor of the extracted coal samples of NMP and ethylenediamine is less than the YM delta factor. Therefore, the Drum factor (g) of NMP and ethylenediamine extraction of coal samples is less than YM.

Figure 4.23 shows the pattern of change of linewidth ΔH and Drum factor (g) before and after YM extraction. Figure 4.23 shows that there is an obvious correlation between linewidth ΔH and Drum factor (g), and the larger the Drum factor g is, the larger the linewidth ΔH is, and the Drum factor (g) is affected by the type and content of groups in the coal molecule as analyzes in the previous subsection. It can be seen that the size of the linewidth of the EPR spectra of coal samples also changes depending on the type and content of the groups in the coal molecules.

As can be seen from Fig. 4.23, the ΔH of the extracted coal samples of ethylenediamine and NMP are smaller than YM, 5.38 and 6.33, respectively, while the ΔH of the extracted coal samples of petroleum ether and toluene (7.49 and 7.61, respectively) are larger than that of YM. Based on the results of the analysis of the g-factor, it can be seen that the percentage of aliphatic and aromatic groups in the extracted coal samples of ethylenediamine and NMP increases compared to that of YM, and the content of groups containing oxygen (O) atoms increases compared to YM for the

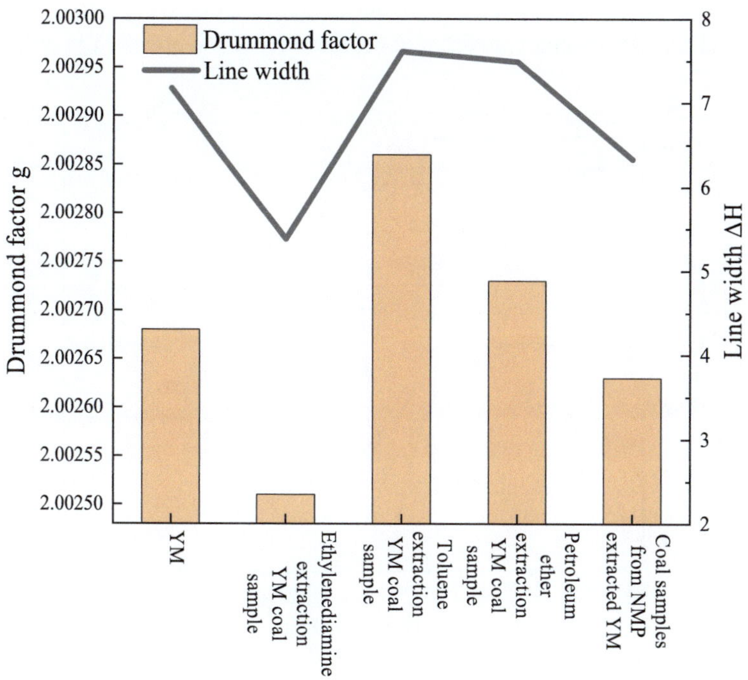

Fig. 4.23 Variation of line width ΔH versus Drum factor g before and after YM extraction

extracted coal samples of petroleum ether and toluene. Petroleum ether and toluene extracted coal samples showed an increase in the content of groups containing oxygen (O) atoms compared with YM. Therefore, according to the physical significance of the linewidth, i.e., the linewidth ΔH is inversely proportional to the relaxation time, it can be seen that the relaxation time of the groups containing oxygen atoms in the coal molecules is shorter. The relaxation time of a group is a reflection of its reactivity, so it can be seen from Fig. 4.5 that the groups containing oxygen atoms in the coal molecule (–OH and carbon–oxygen groups) are more reactive, and the linewidth of the –OH group is the shortest, indicating that it is the most reactive. It can be concluded that –OH is the most dominant reactive group in the low-temperature oxidation stage of coal.

4.3.4 Characteristics of Radical Changes Before and After WYM Extraction

The EPR experimental curves of the coal samples before and after WYM extraction were plotted according to the experimental data, and the changes of the factor (g), line width and radical concentration were investigated separately. As shown in Fig. 4.24, the specific EPR parameters are shown in Table 4.6.

As can be seen from Fig. 4.24, the radical concentration, species and properties in the coal samples changed to different degrees before and after WYM extraction.

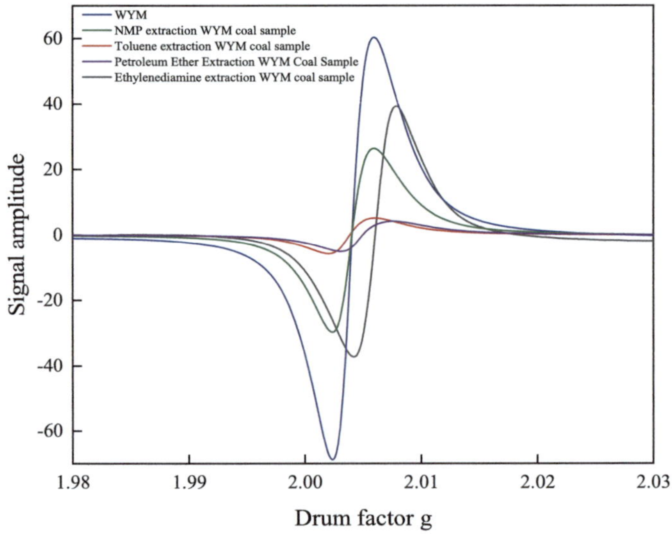

Fig. 4.24 EPR experimental curves of WYM raw coal samples and extracted coal samples

Table 4.6 Parameters of EPR spectra of WYM's extracted coal samples

Sample	Drum factor (g)	Line width ΔH (G)	Radical concentration C (10^{17}/g)
WYM	2.00259	5.9	72.5
Ethylenediamine extraction WYM coal sample	2.00236	4.71	61.78
Toluene extract WYM coal sample	2.00272	7.29	36.47
Petroleum ether extract WYM coal sample	2.00264	6.17	42.39
NMP extract WYM coal sample	2.00243	5.34	59.47

To further assess the reasons for the changes, the calculated Drum factor (g), line width ΔH and free radical concentration C were summarized in Table 4.6:

From the analysis of radical concentration in Table 4.6, it can be seen that the radical concentration C is no longer obviously correlated with the Drum factor (g) and line width ΔH. It seems that the radical concentration C shows a decreasing trend after extraction, but the decreasing amplitude after extraction with different extractants in WYM is quite different from that after extraction with different extractants in YM. The radical concentration in the toluene-extracted coal samples is 36.47 × 10^{17}/g, with a decreasing amplitude of 41.53%, while that in the ethylenediamine extracted samples is 61.78 × 10^{17}/g, with a decrease of 14.78%, which is the same as that in the same extractants in YM. decreased by 14.78%, which is a big difference with the results of the same extractant in YM. This change indicates that toluene extractant extracted more radicals, and ethylenediamine extracted relatively less radicals. And this result and the difference between the extraction effects of ethylenediamine and toluene on –OH groups and aromatic groups, respectively, can be hypothesizes that the content of aromatic radicals in WYM is higher than that of –OH radicals.

From the analysis of Table 4.6, it can be seen that the WYM Drum factor (g) is larger than that of ethylenediamine extracted coal samples and NMP extracted coal samples and smaller than that of toluene extracted coal samples and petroleum ether extracted coal samples, which is the same as the change rule in YM. The reason for this phenomenon is that the residual –OH and carbon–oxygen radicals in the toluene and petroleum ether extracted coal samples are more, i.e., the content of O atoms in the extracted coal samples rises, and the Drum factor (g) of the oxygen-containing radical is larger than that of the C, so the Drum factor g of the extracted coal samples is larger than that of the YM; whereas, in the extracted coal samples of NMP and ethylenediamine, the residual aliphatic and aromatic rings are more, which is equivalent to the increase of the proportion of aliphatic and aromatic groups in the samples, which is similar to that of the coal samples. According to the previous study, the free radical g values of aromatic or chain hydrocarbons with electrons in π-orbitals are 2.0035 and 2.0046, respectively, which are both smaller than the YM Drang factor (g), and thus the Drang factor of the extracted coal samples is smaller

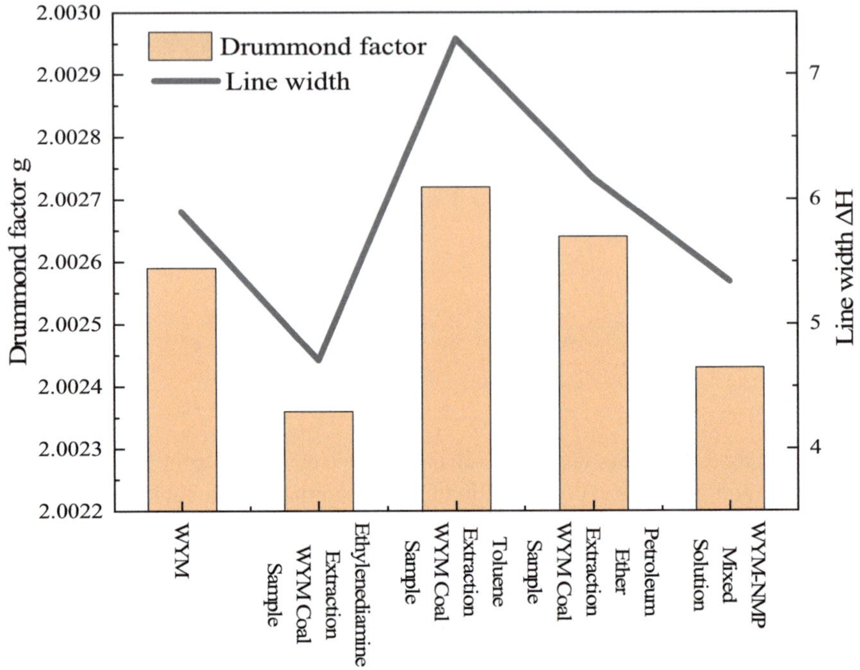

Fig. 4.25 Plot of line width ΔH versus Drum factor (g) before and after YM extraction

than that of the YM. The linewidths of ΔH are plotted together with the Drang factor (g) as shown in Fig. 4.25.

From Fig. 4.25, it can be clearly seen that there is an obvious correlation between the linewidth ΔH and the Drum factor (g). The larger the Drum factor (g), the larger the linewidth, and the reason for this phenomenon is the same as that of the analysis of the YM Drum factor (g) and the linewidth ΔH in the previous subsection, that is to say, we can think that the linewidths of the –OH and the carbon–oxygen radical are larger, and the linewidths of the aliphatic and the aromatic radicals are relatively small, and as the linewidth ΔH is inversely proportional to the relaxation time, it implies that the relaxation times of the –OH and the carbon–oxygen radical are shorter. and inversely proportional to the relaxation time, it means that the relaxation time of –OH and carbon–oxygen radicals is shorter. Combined with the analysis of the relaxation time and reaction difficulty in the previous subsection, it can be concluded that –OH and carbon–oxygen are more reactive, and according to the analysis of the groups before and after extraction in this chapter, combined with the conclusion that the content of the –OH group in the coal sample extracted by WYM toluene is higher than that in the coal sample extracted by WYM petroleum ether, it can be verified that –OH is more effective on coal than in other coal samples, and it can be concluded that –OH is more effective on coal than in other coal samples.

Once again, it is verified that –OH has a greater influence on the coal-oxygen complex reaction in the low-temperature oxidation stage of coal.

4.4 Evolution of Reactive Groups in Spontaneous Combustion of Coal

Reactive groups are the key to the reaction of coal samples, but the FT-IR test can only reflect the change rule of reactive groups before and after the extraction of coal samples, and cannot study the changes of these reactive groups in the spontaneous combustion process of coal. On the basis of the study of coal molecular structure in Chap. 2 and the study of changes in microscopic characteristics of coal samples before and after extraction in this chapter, the changes in the active groups of coal samples before and after extraction were analyzed in the whole coal auto-ignition process by using In-situ FT-IR (Liu et al. 2020) to lay the foundation for the study of the influence of each active group in the coal molecule on the oxidative properties of coal auto-ignition.

4.4.1 Experimental Apparatus and Devices

A Bruker INVENIO S Fourier Transform Infrared (FT-IR) spectrometer was utilized to test for microscopic group changes during the warming process of the coal body, as shown in Fig. 4.26. The principle of using this device is described in Chap. 2, Sect. 4.3 of this paper. In addition, the system is equipped with an in-situ heating cell, a temperature control device, a water cooling system, a gas mixing device, etc., which can realizes the non-destructive in-situ testing of the microscopic groups of the coal body under different atmospheres and temperatures.

4.4.2 Experimental Methodology

In this paper, the in situ diffuse reflectance method technique was chosen to collect and process the coal sample groups during the warming process during the test. The experimental operation process is as follows:

(1) Open the spectrometer and check that the optical path diagram and so on are in normal use. Turn on the temperature control, gas cylinder, gas mixer, etc. to ensure that they work properly.
(2) Put dry potassium bromide powder into the cuvette for testing to collect the background basis vectors. Fill the crucible evenly with the weighed coal sample of about 5 mg, organize it so that the surface is flat and then cover the window

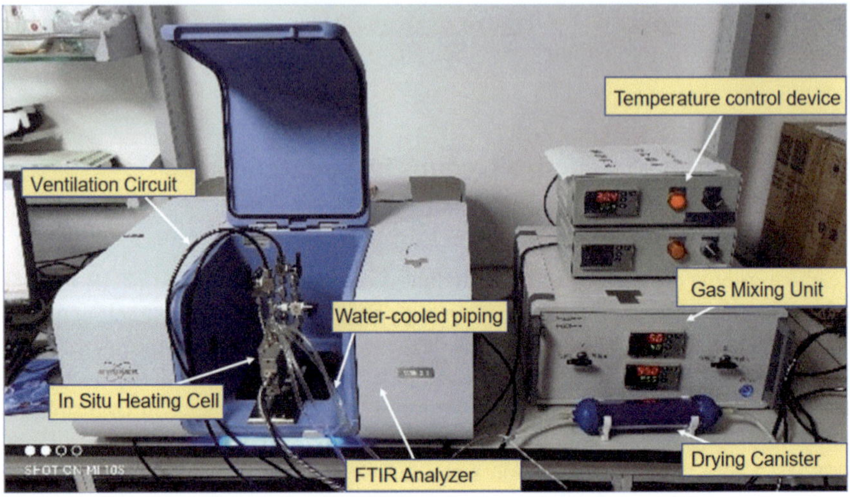

Fig. 4.26 In-situ Fourier transform infrared spectroscopy experimental setup

sheet to seal it for testing, and check the optical path and light intensity to ensure the accuracy of the test.

(3) Set the wave number range of infrared spectrometer as 650–4000 cm^{-1}, the scanning number as 32 times, and the collected signal as absorbance. The grating width is 8 nm, the scanning frequency is 7.5 kHz, and the scanning mode is continuous scanning. The temperature control device was set to increase the temperature range from 30 to 300 °C, the temperature increase rate was 10 °C/min, and the total gas flow rate was set to 50 mL/min.
(4) At the end of the experiment, remove the coal samples, clean the in-situ cell well and carry out the next set of experiments.

4.4.3 Analysis of Experimental Results

The infrared spectra of YM and WYM raw coals obtained from the tests are shown in Fig. 4.27. The results showed that the absorbance of various different microscopic groups in the coal samples had different trends as the temperature increased. For example, the –OH group of the two coals with a wave number of roughly 3400 cm^{-1} shows an extremely obvious decreasing trend before 100 °C, while the decreasing trend of the absorbance of the –C=C– group with a wave number of 1600 cm^{-1} in this temperature range is relatively weak. This indicates that different microscopic groups have different reaction histories during the whole reaction process. In addition, it can be observed that the infrared curves of YM and WYM are substcnticlly different, which implies that the degree of metamorphism has a significant effect on

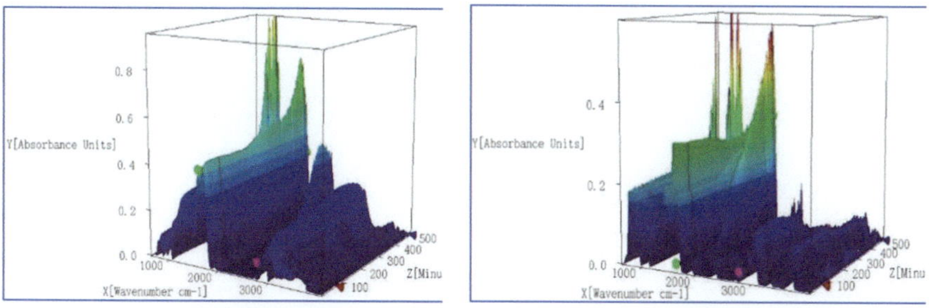

(a) YM (b) WYM

Fig. 4.27 Characteristics of infrared curve changes during oxidation and warming of coal samples

the evolution of coal microgroups. Therefore, to clarify the influence of coal oxidation microcourse and extraction on the evolution of coal microstructure, the trend of coal medium groups was analyzed in detail based on the results obtained from the above analysis.

(a) **Hydroxyl**

The activity of hydroxyl group (–OH) as one of the most active types of microscopic groups in coal has attracted much attention. It is widely recognizs by many scholars that the hydroxyl group plays a crucial role in the early stages of coal combustion reactions (Li et al. 2017; Lu and Wang 2022). Figures 4.28 and 4.29 give the trends of two different hydroxyl groups in coal, phenolic or alcoholic hydroxyl (–OH–1), and conjugated hydroxyl (–OH–2), respectively. The results show that both hydroxyl groups showed a decreasing trend during coal combustion but their evolution stages are different. At the initial stage of coal combustion, with the evaporation of water from the coal, the bonding of the coal is rapidly broken, which causes a rapid decrease of –OH–2 in the coal before 100 °C. After the evaporation of free water in coal was completed, the bound water in coal commenced to gradually precipitate and evaporate, which caused a second decrease in –OH–2. In contrast, the decrease of –OH–1 in coal comes from the phenolic hydroxyl group on the benzene ring and the alcohol hydroxyl group on the aliphatic hydrocarbons in coal. The rate of oxidation of the two to peroxide is small in the pre-combustion reaction at lower temperatures therefore the decline is slower. When the temperature exceeds ca. 300 °C, the volatile components in the coal start to burn vigorously, and the large amount of heat generated promotes the oxidation of phenolic hydroxyls and alcohol hydroxyls in the coal, leading to a further decrease of both. Comparing the trends of the two hydroxyl groups in YM and WYM, it can be found that the decreasing rates of both –OH–1 and –OH–2 in WYM are weaker than that in YM, the former is because the stability of WYM is stronger than that of YM, so its alcohol hydroxyl group has lower reactivity, and the latter is because there is less water in WYM, so its water evaporation effect is weak, leading to a weaker decreasing trend of its –OH–1 weaker decreasing trend.

Comparing the trends of –OH changes between the raw coal and the extracted coal samples, it can be found that the extractant has different extraction effects on

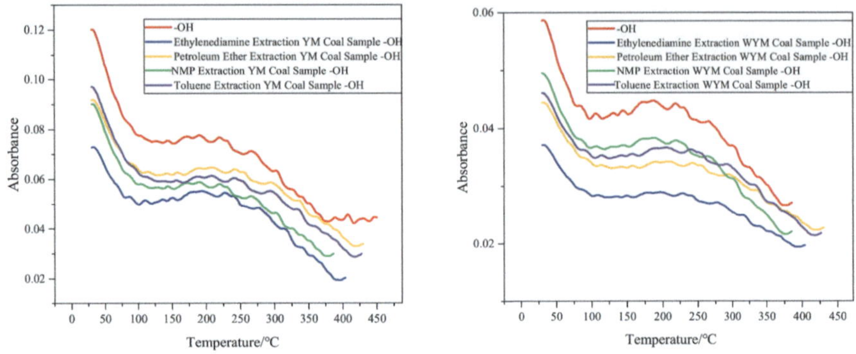

(a)YM (b)WYM

Fig. 4.28 Trend of –OH (3400 cm^{-1}) for two coal samples

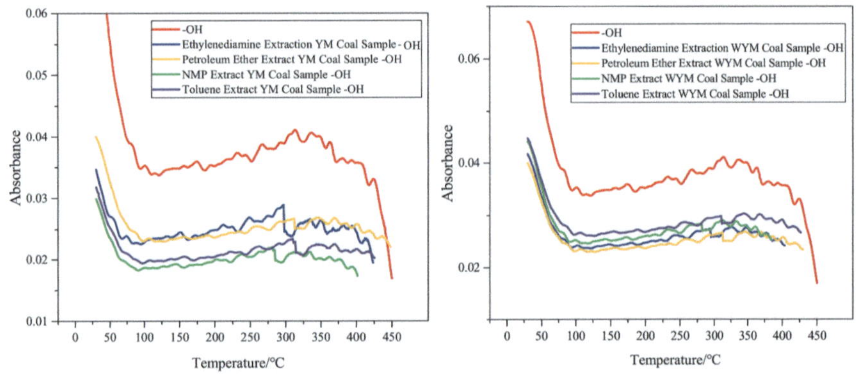

(a)YM (b)WYM

Fig. 4.29 Trend of –OH (3200 cm^{-1}) for two coal samples

two different types of hydroxyl groups. Among them, the decreasing trend of –OH–2 in the water evaporation stage of all four extracted coal samples was prominently smaller than that of the original coal. This is mainly because the coal samples need to be dried after extraction, which leads to the reduction of water in the extracted coal samples and then reduces the water evaporation effect, so the decreasing trend is weaker. For –OH–1, the decreasing trend of ethylenediamine-extracted coal samples was weaker than that of other coal samples because hydroxyl group is one of the most critical groups in the spontaneous combustion process of coal. The results in Chap. 3 showed that ethylenediamine has a better extraction effect on the hydroxyl group in coal, so the –OH–1 in coal samples decreased after extraction. At the same time, the decrease of –OH–1 led to a certain degree of decrease in the reactivity of the coal, so its decreasing trend was weakened.

(b) Aromatic Hydrocarbon Compounds

Aromatic hydrocarbons are the backbone of the coal structure, and their stability is relatively high during the oxidation process. Figure 4.30 Trends of –C=C– ($1600\ cm^{-1}$) in the spontaneous combustion process before and after extraction of two coal samples. The results show that with the increase of temperature, the raw coal and extracted coal samples of both YM and WYM show a sclient decreasing trend. This implies that the aromatic ring structures of both coal samples were in the process of slow oxidative cleavage during the whole spontaneous combustion process. Further, it can be observed that the decreasing trend of –C=C– is characterizs by obvious stages. Among them, the decreasing trend of the curves is obviously enhanced at about 350 °C for YM and 400 °C for WYM, which implies that the aromatic ring cleavage tendency of YM and WYM is accelerated at 350 and 400 °C, respectively. Obviously, at the beginning of the reaction, owing to the lower temperature, the reaction intensity between coal and oxygen is smaller therefore the cracking rate is slower. When the reaction temperature exceeded the ignition point, the cleavage of the aromatic ring structure accelerated, and a large amount of volatile matter and carbon oxides were generated, accompanied by intense exothermic, and the coal samples began to burn violently. The aromatic ring of WYM has a higher degree of condensation, stronger stability, and weaker tendency to spontaneous combustion, so the rapid cleavage temperature of its aromatic ring is significantly greater than that of YM.

Comparing the trends of –C=C– of the two coal samples before and after extraction with different extractants, it can be found that the trends of –C=C– of the coal samples before and after extraction did not change prominently, which means that the extraction did not change the physicochemical properties of the aromatic hydrocarbon compounds in the coal, and it proves that the extraction technique is a physical effect on the coal without involving the evolution of the radicals. This means that

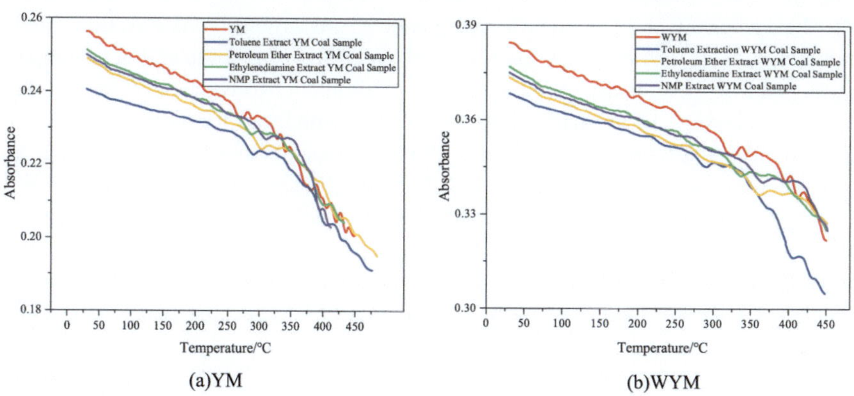

Fig. 4.30 Trend of –C=C– for two coal samples

the extraction did not change the physicochemical properties of the aromatic hydrocarbon compounds in the coal, which proves that the extraction technique has a physical effect on the coal and does not involve the evolution of the groups. It was further found that the decreasing slopes of the –C=C– curves of the toluene extracted coal samples were slightly smaller than those of the other samples, especially in the slow stage, and the decreasing slopes of the two toluene extracted samples were substcnticlly smaller than the other curves, which implies that the decomposition of the aromatic rings of the coal samples before combustion after toluene extraction has a weaker tendency. From the conclusions in Chap. 3, it can be seen that toluene has a better extraction effect on aromatic hydrocarbon compounds in coal, so the unstable part of aromatic hydrocarbons or benzene ring fragments in the two types of coals after toluene extraction were extracted, and the continuous decomposition of the latter two in the coal before combustion is one of the key factors in the –C=C–, so the decreasing trend of –C=C– of the two kinds of toluene extracted samples before combustion is weaker than that of the other curves. –C=C– decreasing trend is weakened.

(c) **Aliphatic Hydrocarbon Compounds**

Aliphatic hydrocarbons play a crucial role in the structure of coal, closely connecting the aromatic ring cores in coal and acting as a bridge between the cores (Qi et al. 2018). Figure 4.31 gives the change process of two key aliphatic hydrocarbon compounds, methyl (–CH_3) and methylene (–CH_2–), during spontaneous combustion of coal samples after extraction with different extractants. The results in Fig. 4.32 show that both aliphatic hydrocarbon compounds in the coal samples before and after extraction began to rise gradually with increasing temperature, which means that at this time the rate of generation of aliphatic hydrocarbons in coal is greater than the rate of consumption. Combined with the trend of aromatic hydrocarbons, it can be seen that the rise of aliphatic hydrocarbons in coal is due to the breakage of molecular bonds after the coal is heated. On the one hand, the structural stability of coal deteriorates after being heated, and the long-chain aliphatic hydrocarbons break into short-chain and release hydrocarbon gases, such as C_2H_4 and C_2H_6 under the action of temperature, resulting in an increase in the content of aliphatic hydrocarbons. On the other hand, the cleavage of aromatic rings in coal also enhanced the generation rate of aliphatic hydrocarbons. However, the aliphatic hydrocarbons in coal began to gradually decrease with further increase in temperature, where the turning point of YM occurred at about temperature of roughly 200 °C, while WYM occurred at approximately 250 °C. This indicates that the aliphatic hydrocarbons in the coal begin to be gradually oxidized and consumed at elevated temperatures. The temperature at which WYM begins to decline is greater than that of YM because of its more stable nature.

Observing the decreasing trend of the two types of aliphatic hydrocarbons in the raw coal and the extracted coal samples, it can be found that the trend of the aliphatic hydrocarbons in the coal samples before and after extraction did not change scliently, which is basically consistent with the conclusion of the aromatic hydrocarbons, i.e., the extraction effect of the extractant is mainly by physical rather than

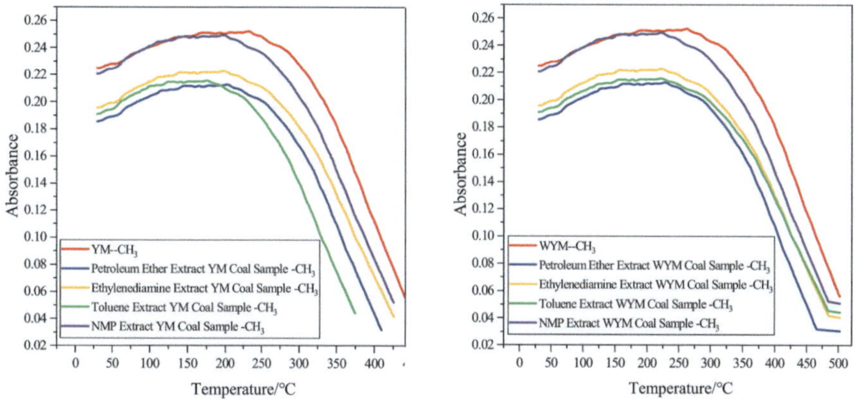

(a)YM (b)WYM

Fig. 4.31 Trend of different coal samples $-CH_3$

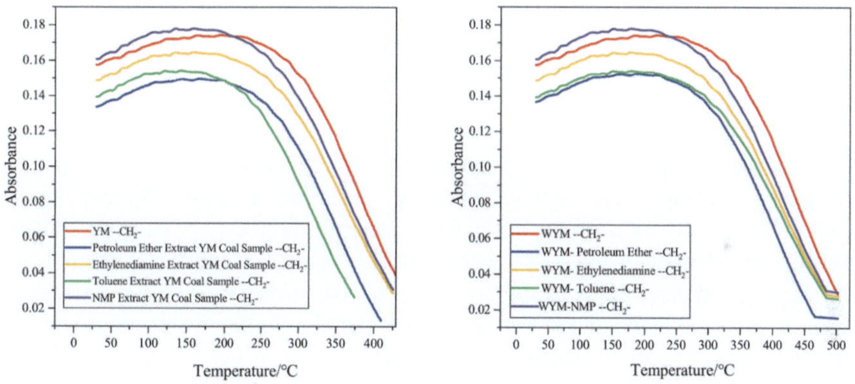

(a)YM (b)WYM

Fig. 4.32 Trend of $-CH_2-$ for different coal samples

chemical effects. Among them, the trends of the two aliphatic hydrocarbons in the petroleum ether extracted coal samples were somewhat different from those of other coal samples. With the increase of temperature, the growth trend of the aliphatic hydrocarbons of the petroleum ether extracted coal samples was weaker than that of the other coal samples, which indicated that part of the long-chain aliphatic hydrocarbons and more aromatic ring groups were taken away from the coal after petroleum ether extraction, resulting in the decrease of the aliphatic hydrocarbons generation rate at the early stage of heating, which, in turn, affected the reactivity of the reaction. In the later stage of the reaction, the consumption of aliphatic hydrocarbons is mainly due to the oxidation of oxygen, so the trends are basically the same.

(d) **Carbon–oxygen Groups**

Carbon oxides are a vital part of the carbon–oxygen groups in coal, and they are widely recognized as one of the direct sources of CO and CO_2 produced by coal combustion (Menon and Patnaikuni 2017). The trends of the three carbon–oxygen compounds after YM and WYM combustion are shown in Figs. 4.33, 4.34, and 4.35 for the carbon–oxygen single bonds (–C–O–), carbonyl group (–C=O) and carboxyl group (–COO–). The results in Fig. 4.34 show that the three carbon–oxygen groups show different trends with increasing temperature. Among them, the change of –C–O– is more complicated throughout the combustion process. The content of –C–O– began to increase with increasing temperature, indicating that the rate of –C–O– generation in coal at the early stage of the reaction was greater than its consumption rate. After the temperature exceeds about 100 °C, the –C–O– in the coal begins to decrease slightly, with the decrease in YM lasting until roughly 300 °C and that in WYM until approximately 350 °C or so.

Observing the trends of –C=O and –COO–, it can be found that the contents of the two groups first remain basically unchanged and then rise rapidly as the reaction proceeds, in which the –C=O and –COO– were rising at a transition temperature of about 250 °C for YM and about 300 °C for WYM. This is a prominent correlation with the turning trend of –C–O– and aliphatic hydrocarbons in the previous section. In the early stage of spontaneous combustion of coal, the gradual decomposition of –C=O and –COO– in coal into gaseous products, such as CO and CO_2, volatilizs by heat leads to the depletion of their contents, but at the same time they are replenished by the –C–O group will be replenished by the oxidation of the –C–O group, so the content is basically stable in the initial stage. After the temperature exceeds 250 and 300 °C, the intensity of the coal oxygen complex increases, and the aromatic rings and aliphatic hydrocarbons of the coal begin to be oxidised in large quantities to produce carbon oxides, and at the same time, the elevated temperature provides a convenient way to further oxidize –C–O–, which leads to the rapid decline of –C–O–.

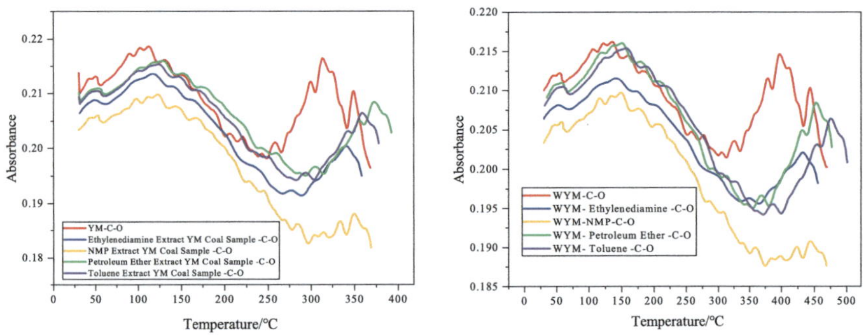

(a)YM (b)WYM

Fig. 4.33 Trends of different coal samples –C–O–

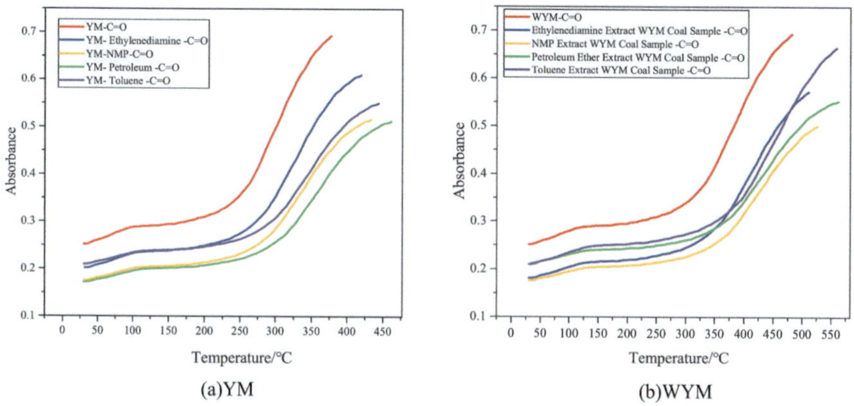

Fig. 4.34 Trends of different coal samples –C=O

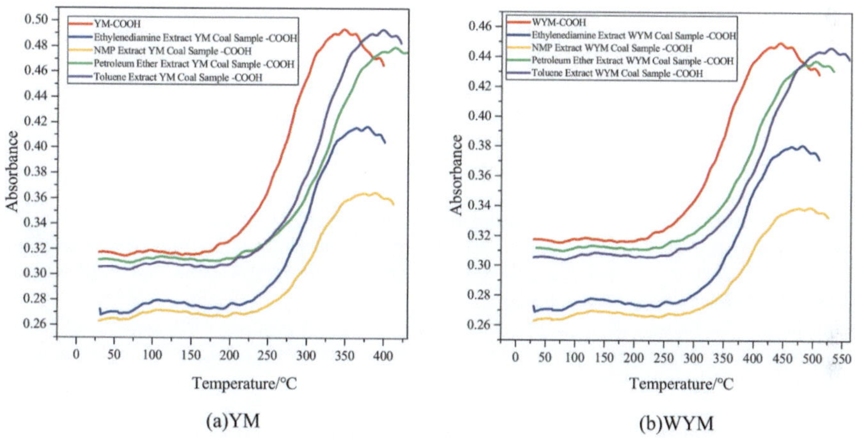

Fig. 4.35 Trend of different coal samples –COO–

rapid decrease of –C–O– and rapid increase of –C=O and –COO– content. Similar to the above analysis, the enhanced molecular stability of WYM leads to a lag in the reaction of its radicals compared to YM, which, in turn, leads to an increase in the transition temperature.

Comparing the trends of the three carbon–oxygen groups of the original coal and the extracted coal samples with different extractants, it can be found that, similar to the results above, the extraction did not change the oxidation reaction course of the groups in the coal molecules, but the trends of the carbon–oxygen groups of the extracted coal samples differed from those of the original coal, with the most obvious difference in the NMP extracted coal samples. For example, the turning points of –C–O– group changes in the original YM coal were about 100 and 250 °C, while those in the NMP extracted coal samples were 120 and 300 °C, respectively. were

120 and 300 °C. This implies that the NMP extractant has a better extraction effect on the three carbon–oxygen groups in the coal, confirming the conclusions in Chap. 3.

References

Dai CW, Li HN, Wang JJ et al (2022) Effect and evaluation of different organic solvents on extraction of Lingwu bituminous coal. Ningxia Eng Technol 21(02):127–132. https://doi.org/10.3969/j.issn.1671-7244.2022.02.007

Fu XY, Li YF, Li YH et al (2024) Efficacy study on the degradation of BPA by activation of persulfate by tributyl phosphate modified biochar. Ind Water Treat 12:1–14

Guo S (2022) Study on extracting coal tar residue and its CO_2 absorption by switchable solvents. Shanxi. https://doi.org/10.27352/d.cnki.gylgu.2022.001883

Guo Z, Li X, Li ZY et al (2023) Research progress on degradative solvent extraction of low-rank coals. Coal Sci Technol 51(06):286–303

Li L, Fan H, Hu H (2017) Distribution of hydroxyl group in coal structure: a theoretical investigation. Fuel 189:195–202. https://doi.org/10.1016/j.fuel.2016.10.091

Liu Y, Nan C, Jia P et al (2020) In-situ FT-IR spectroscopy investigation of CH_4 and CO_2 reaction. Catalysts 10(1):131. https://doi.org/10.3390/catal10010131

Liu QJ (2022) Study on multi-field coupling enhancement of high sulfur coal desulfurization. Chongqing. https://doi.org/10.27670/d.cnki.gcqdu.2022.004118

Lu X, Wang J (2022) Reaction mechanisms involving the hydroxyl radical in the low-temperature oxidation of coal. Fuel 314:122732. https://doi.org/10.1016/j.fuel.2021.122732

Menon K, Patnaikuni V (2017) CFD simulation of fuel reactor for chemical looping combustion of Indian coal. Fuel 203:90–101. https://doi.org/10.1016/j.fuel.2017.04.084

Qi X, Chen L, Ji Y et al (2018) Reaction mechanism and thermodynamic properties of aliphatic hydrocarbon groups during coal self-heating. Energy Fuels 32(10):10469–10477. https://doi.org/10.1021/acs.energyfuels.8b02165

Schmallegger M, Gescheidt G (2018) Antioxidant activity of beer: an EPR experiment for an undergraduate physical-chemistry laboratory. J Chem Educ 95(11):2013–2016. https://doi.org/10.1021/acs.jchemed.8b00329

Xue WH, Du ML, Liu ZC et al (2022) Effects of extraction methods on molecular structure characteristics of Huangling coal. Clean Coal Technol 28(07):96–102

Chapter 5
Oxidation Characteristics of Spontaneous Combustion Active Groups Before and After Coal Extraction

Abstract In this chapter, the characteristics of weight loss, characteristic temperatures, and thermal effects of coal samples before and after extraction were studied comprehensively by thermogravimetric analysis (TG), differential scanning calorimetry (DSC) and temperature-programmed experiments. Then, oxidation characteristics of active groups on coal molecular were investigated symmetrically and comprehensively. Also, their contributions to the occurrence and development of coal spontaneous combustion were clarified. This provides a scientific basis for revealing the mechanism and optimizing prevention and control technology of coal spontaneous combustion.

Coal spontaneous combustion is mainly caused by the reaction of the active groups in the coal molecules with oxygen and the continuous accumulation of therm. This process has obvious characteristics of macroscopic thermal effects, such as characteristic temperature (Yang et al. 2023), heat release rate (Song et al. 2021) and heat release amount (Shu et al. 2023). On the basis of Chap. 4, this chapter conducts oxidative thermograin-mass tests on coal samples before and after extraction (Deng et al. 2017), so as to compare and analyze the thermal effect characteristics of coal samples before and after extraction, and finally provide a basis for explaining the oxidation reaction mechanism of different active groups in the process of spontaneous combustion of coal.

5.1 Experimental Conditions and Methods

5.1.1 Purpose of the Experiment

To further analyze the influence of different active groups on the coal oxidation process, the German (NETZSCH) STA449F3 synchronous thermal analyzer was used for thermal analysis of YM, WYM and their raffinized coal samples (Fig. 5.1).

J. Deng et al., *Coal Spontaneous Combustion Theory with Chain Self-promoted Oxidizing Induced by Active Group*, https://doi.org/10.1007/978-981-96-5362-1_5

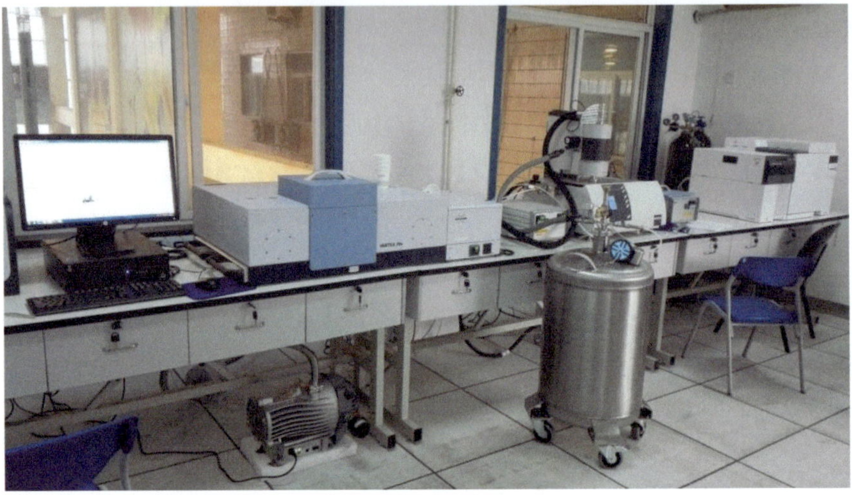

Fig. 5.1 The appearance of the experimental equipment for thermal analysis

By controlling the heating rate, the process information of oxidation reaction of different coal samples was received. The influence of coal metamorphism degree on coal oxidation process was acquired by comparing the difference of YM and WYM thermal analysis curves. The role of different active groups in coal oxidation process was revealed by comprehensive analysis of thermal analysis curves of extracted coal samples with different extractants, and the influence of different extractants on the difficulty of coal oxidation reaction was directly understood.

5.1.2 Experimental Conditions

YM, WYM, and their extractive coal samples used in the experiment were roughly carefully screened for 200 mesh first, and the amount of coal samples placed in each experiment was 15 mg. The starting temperature was 30 °C and the ending temperature was 800 °C. The heating rate was 10 °C/min, the total gas flow rate was 100 mL/min, and the oxygen concentration was 21%. To ensure full contact between the coal sample and oxygen, the crucible cover was not used in the sample crucible and the control crucible.

5.2 Mass and Characteristic Temperature Change Characteristics of Coal Samples Before and After Extraction

5.2.1 *Coal Oxidation Process Analysis*

(a) Variation Characteristics of TG-DTG Curves of Coal Samples

By analyzing the thermogravimetric test results of YM, WYM and their extract coal samples, it is found the quality of different coal samples varies with temperature in the process of heat oxidation. According to the different trends of coal samples during the heating process, it can be divided into four stages, namely, the mass loss stage of water volatilization and gas desorption, the mass gain stage of oxygen absorption and thermal decomposition stage, the combustion stage and the burnout stage. Taking the TG curve of YM raw coal sample as an example, the changes in different stages during the experiment were analyzed, as shown in Fig. 5.2.

It can be seen from Fig. 5.2 that the moisture contained in the coal sample gradually volatilized with the increase of temperature, and during this process, the gas contained in the pores of the coal sample also desorbed with the increase of temperature, and the original adsorption/desorption balance of the coal sample was broken, resulting in mass loss of the coal sample. With the further increase of temperature, the active groups in the molecular structure of coal are gradually activated, and the adsorption

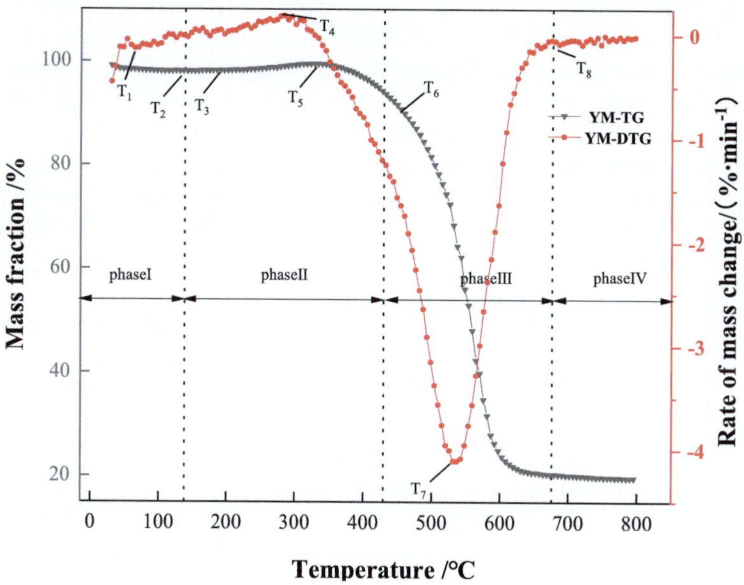

Fig. 5.2 TG/DTG curve of YM raw coal oxidation process

capacity of coal sample for oxygen is gradually enhanced (Jia et al. 2024). At this time, both physical adsorption and chemisorption exist, but the mass of adsorbed oxygen is still smaller than the mass of gas desorption. At this time, the mass loss rate of coal sample slows down until the oxygen adsorption capacity and gas desorption capacity reach a dynamic equilibrium. The quality of coal sample remains basically unchanged. As the temperature continues to rise, the number of oxygen-absorbing active groups in coal increases sharply, and the rate of coal-oxygen recombination reaction continues to increase (Hao et al. 2022). The dynamic equilibrium state is gradually broken. At this time, the mass of adsorbed oxygen in the coal sample is greater than the mass of gas released by the coal-oxygen recombination reaction, and the coal sample begins to demonstrate a trend of oxygen-absorbing mass gain. Until the mass of the adsorbed oxygen and the mass of the released gas reach a new equilibrium state, the mass gain of the coal sample reaches the maximum. Then the coal-oxygen recombination reaction rate further increases, the mass of the released gas is greater than the mass of the adsorbed oxygen, and the coal sample enters the violent decomposition stage. When the temperature reaches the ignition point, the aromatic ring and other structures with strong stability in the coal molecules are expeditiously oxidized and decomposed, and the oxidation rate of the coal sample surges and starts to burn, producing an enormous amount of gases (CO, CO_2, CH_4, C_2H_4, C_2H_6, etc.). The promptly of the coal sample decreases rapidly until the organic matter in the coal sample is exhausted and the quality no longer changes.

(b) **Characteristic Temperature Point**

Coal has obvious stages in the oxidation process. Taking YM as an example, its characteristic temperature points can be divided into T_1 (critical temperature), T_2 (cracking temperature), T_3 (active temperature), T_4 (growth temperature), T_5 (maximum mass temperature), T_6 (ignition point temperature), T_7 (maximum mass loss rate temperature) and T_8 (burnout temperature) according to the change characteristics of TG curve (Fig. 5.2). As a vital parameter in the coal oxidation process, the value of the characteristic temperature point represents the difficulty of coal sample oxidation to a certain extent, and is of great significance for the analysis of spontaneous combustion of different coal samples (Wang 2005, 2015) (Tables 5.1 and 5.2).

5.2.2 Comparative Analysis of TG Curves of YM and Its Raffinate Coal Samples

To appraise analyze the influence of different groups on the YM oxidation reaction process, thermogravimetric analysis was carried out on YM coal samples before and after extraction, and the results are shown in Fig. 5.3. YM raw coal sample and its extractive coal sample showed different degrees of mass loss before 150 °C. This is mainly due to desorption and volatilization of water in coal and gas in pore structure with the increase of temperature (Bai et al. 2017). As the temperature continues to rise,

Table 5.1 Characteristic temperature points of YM and its raffinate coal sample

Sample	T_1	T_2	T_3	T_4	T_5	T_6	T_7	T_8
YM	68.95	135.48	203.81	255.87	302.85	429.32	505.01	616.18
Ethylenediamine extract YM coal sample	69.42	116.17	185.28	246.10	299.16	467.46	562.44	681.21
NMP extract YM coal sample	70.72	119.42	186.97	238.68	295.53	459.7	532.29	677.67
Petroleum ether extract YM coal sample	60.68	124.86	195.73	244.75	295.97	449.91	538.78	656.89
Toluene extract YM coal sample	76.84	128.76	199.77	252.50	286.77	431.01	511.07	665.27

Table 5.2 Characteristic temperature points of WYM and its raffinate coal samples

Sample	T_1	T_2	T_3	T_4	T_5	T_6	T_7	T_8
WYM	78.81	124.14	240.13	305.09	409.67	518.78	605.38	718.23
Ethylenediamine extract WYM coal sample	104.08	137.54	219.52	295.60	394.11	520.71	620.77	729.66
NMP extract WYM coal sample	91.62	128.47	228.20	299.96	391.24	526.64	616.09	731.17
Petroleum ether extract WYM coal sample	70.53	117.19	223.77	296.39	396.34	529.86	618.12	726.25
Toluene extract WYM coal sample	87.87	133.01	234.86	303.16	391.66	523.32	614.26	721.88

part of the active groups in the coal are activated, and the oxygen absorption capacity of the coal sample is gradually enhanced. At this time, the physical adsorption and chemical adsorption of oxygen exist simultaneously, the mass gain of coal-oxygen reaction and the mass loss of gas volatilization exist simultaneously, and the oxygen quality of coal maintains a dynamic equilibrium state.

Between 200 and 300 °C, YM raw coal and its extract coal samples both show a mass gain trend. In this stage, oxygen absorption mass gain of coal and desorption mass loss of gas generated by coal-oxygen reaction exist simultaneously, but the mass gain of coal-oxygen composite reaction plays a dominant role, so it shows a mass gain trend. For 300–600 °C, YM raw coal and its extract coal samples show a trend of swift mass loss, which is mainly due to the rise in temperature, which leads to the decomposition reaction of coal samples, decomposing a large number of combustible substances into gas, and eventually the quality of coal samples is swift decreased.

To further study the effect of ethylenediamine extraction on the oxidation process of YM coal sample, YM and ethylenediamine extraction coal sample were compared and analyzed. As can be seen from Fig. 5.4a, there is a salient difference in quality loss between YM raw coal sample and ethylenediamine raffinate coal sample from the initial temperature to 150 °C, and YM raw coal has a greater mass loss than

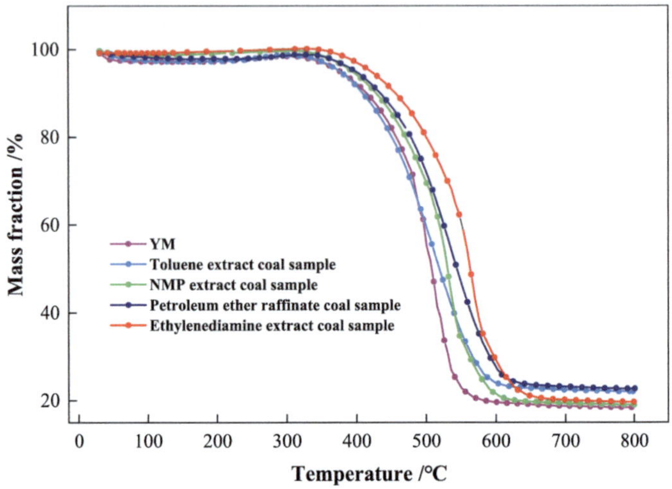

Fig. 5.3 TG curve of YM and its raffinate coal sample during oxidation process

ethylenediamine raffinate coal sample. As can be seen from the analysis in the above section, because the content of –OH in the coal sample is less, the reaction rate of coal and oxygen is limited. According the amount of desorption gas can be generated less, and the coal sample can more achieve the balance between oxygen absorption mass gain and gas desorption mass loss.

However, between 300 and 650 °C, a large number of active groups (such as –CHO and –COOH, etc.) are accumulated in the coal sample during the early reaction. With

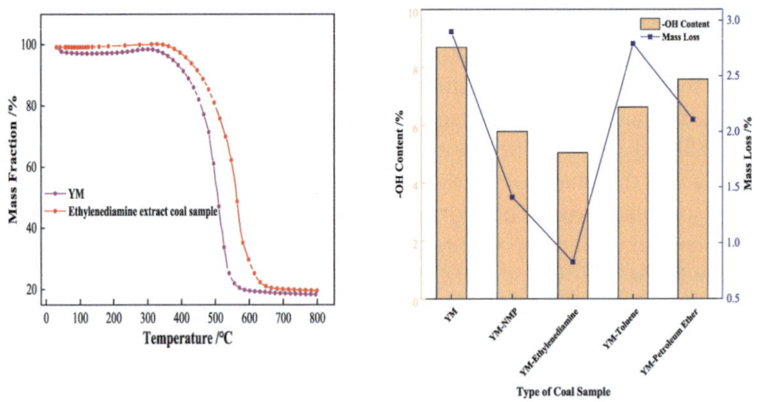

(a) Comparison of TG curves of YM and ethylenediamine raffinate YM coal samples

(b) The relationship between −OH content and quality loss of coal samples

Fig. 5.4 Comparison of TG curves of YM and ethylenediamine raffinate YM coal samples and the relationship between –OH content and quality loss of coal samples the further increase of temperature, these active groups begin to undergo a violent oxidation reaction with oxidation, producing a large number of gas products, such as CO_2, CO, which promotes the oxidation process of coal. Compared with YM raw coal sample, the temperature in the swift mass loss interval of ethylenediamine raffinate coal sample is improved, which is because the content of the –OH active group in the ethylenediamine raffinate coal sample is less than that of YM raw coal sample, and –OH is a key factor to promote the combustion and decomposition of aliphatic groups in coal. Therefore, the decrease of –OH in the ethylenediamine raffinate coal sample leads to the need for higher temperature before combustion reaction can occur, and the whole combustion reaction interval shows hysteresis.

Based on the analysis of group proportion in coal samples in Chap. 4, the –OH group proportion and mass loss at 130 °C of YM and its four raffinate coal samples were compared and analyzed, as shown in Fig. 5.4b. It can be seen from Fig. 5.4b that the quality loss of YM and its raffinate coal samples is prominently correlated with the –OH content. With the decrease of –OH content, the quality loss of coal samples also gradually decreases. Compared with YM raw coal sample, the –OH content of ethylenediamine raffinate coal sample is the least, and –OH is a vital active group to initiate numerous elementary reactions in the combustion process. It can be intuitively seen from Fig. 5.4b that the mass loss of ethylenediamine raffinate coal sample is smaller than that of YM original coal sample and other raffinate coal samples. Therefore, it can be analyzed that the reduction of –OH content in ethylenediamine raffinate coal sample appraised the intensity of coal-oxygen recombination reaction, resulting in smaller gas desorption generated by the reaction between adsorbed oxygen and active groups. This further corroborate that the amount of –OH in the coal sample will affect the reaction rate of the coal-oxygen recombination reaction. The higher the amount of –OH in the coal sample, the faster the reaction rate of oxygen, the greater the mass loss ratio of the coal sample; conversely, the lower the amount of –OH in the coal sample, the lower the coal-oxygen recombination reaction rate, and the smaller the mass loss ratio of the coal sample.

In addition, between 300 and 600 °C, when the oxidation of the coal sample enters the high temperature stage, a large number of active groups in the coal participate in the oxidation reaction, which promotes the oxidation process of the coal sample, and the oxidation reaction occurs promptly in the coal sample within the range, producing an immense amount of gas, while the coal sample itself shows rapid mass loss. Therefore, when the combustion rate of petroleum ether raffinate coal sample and NMP raffinate coal sample is maximum, the corresponding temperature is higher than that of YM raw coal sample because the number of active groups that can promote coal oxidation in raw coal sample is greater than that of petroleum ether raffinate coal sample and NMP raffinate coal sample, so it can be burned at a lower temperature and reach the temperature point of maximum combustion rate faster. To further analyze the effect of toluene extraction of aromatic hydrocarbons from YM on the oxidation process of YM, YM and toluene raffinate coal samples were compared and analyzed, as shown in Fig. 5.5b. As can be seen from the figure, the curve changes of toluene extract coal sample and YM original coal sample are basically the same before 300 °C, which is mainly because the aromatic groups in

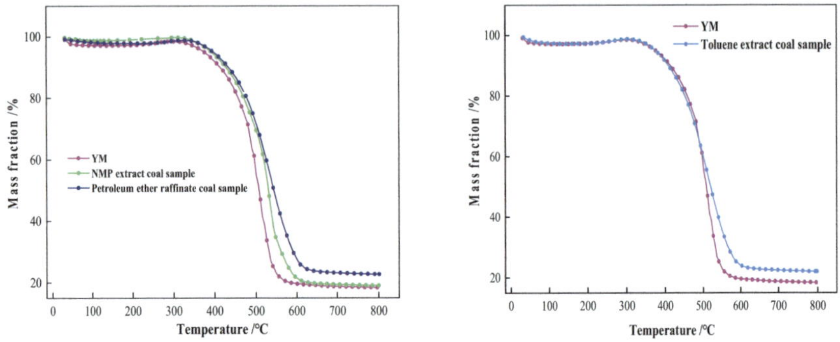

(a) YM, petroleum ether extract YM coal samples and toluene extract YM coal samples

(b) YM and toluene extract YM coal samples

Fig. 5.5 Comparison and analysis of TG curve of coal sample oxidation process

YM are mainly targeted in the process of toluene extraction, and the key active groups that affect the oxygen adsorption capacity and gas desorption capacity of coal sample before 300 °C are various oxygen-containing groups. However, the aromatic groups extracted by toluene basically do not react at low temperature, so the TG curves of toluene raffinated coal samples are similar to those of YM raw coal samples. However, between 300 and 600 °C, the TG curves of toluene extract coal sample and YM original coal sample have slightly changed, but the overall trend is still basically the same, mainly because the aliphatic group mainly plays a key role in the initial stage of coal decomposition and combustion, and toluene has basically no extraction effect on the aliphatic group. Therefore, there is no distinct change in TG curves of decomposition and combustion in the initial stage.

5.2.3 Comparative Analysis of DTG Curves of YM and Its Extracted Coal Samples

Doing first-order differentiation on the TG curve of the coal sample can obtain the DTG curve of the coal sample, which is a parameter reflecting the trend of fast and slow change of the quality of the coal sample in the process of warming. A value >0 on the DTG curve indicates an increase in the mass of the coal sample, while a value <0 indicates a decrease in the mass of the coal sample, and the greater the absolute value of the value, the greater the rate of change.

The DTG data of YM and ethylenediamine extracted coal samples were compared and analyzed, and the results are shown in Fig. 5.6. By comparing the DTG curves of YM and ethylenediamine extracted coal samples, the influence of –OH content on the oxidation reaction rate of coal samples can be understood. On the DTG curve, the position where the absolute maximum value of the mass change rate of the coal samples is located is the maximum peak and valley of the DTG curve, and the absolute value of the mass change rate of the YM and ethylenediamine extracted coal samples is the largest in the peak and valley, so the oxidation reaction of the coal samples is the most intense at the temperature of the corresponding point. The temperature corresponding to the peak and valley of the ethylenediamine-extracted coal samples was higher than that of the peak and valley of the YM coal samples, which indicated that it was more difficult for the ethylenediamine-extracted coal samples to reach the maximum of the reaction rate than that of the YM samples in the oxidation reaction process, due to the fact that after the extraction of YM by ethylenediamine, the amount of –OH contained in the ethylenediamine-extracted coal samples was reduced, and the –OH is a key reactive group for the promotion of the reactions of the numerous radicals of the coal-oxygen composite reaction. The oxidation process of the ethylenediamine-extracted coal samples was slower than that of the raw YM coal samples.

Comparison of the DTG curves of YM, petroleum ether-extracted YM coal samples, and NMP-extracted coal samples can provide insights into the influence of aliphatic and carbon–oxygen groups on the oxidation reaction rate of coal samples. As can be observed from Fig. 5.7a, there are three peaks and valleys with obvious differences in the graph. The temperature points corresponding to the peaks and valleys of the petroleum ether-extracted YM and NMP coal samples are after the temperature points corresponding to the peaks and valleys of the YM coal sample, indicating that the petroleum ether-extracted YM and NMP coal samples need higher temperatures to reach the maximum of the reaction rate during combustion, mainly due to the significantly reduced number of aliphatic and oxygenated reactive groups in the petroleum ether-extracted YM and NMP coal samples as compared with the YM original coal sample, and the above reactive groups have a substantial reduced number of aliphatic and carbonyl-oxygenated reactive groups, while the number of

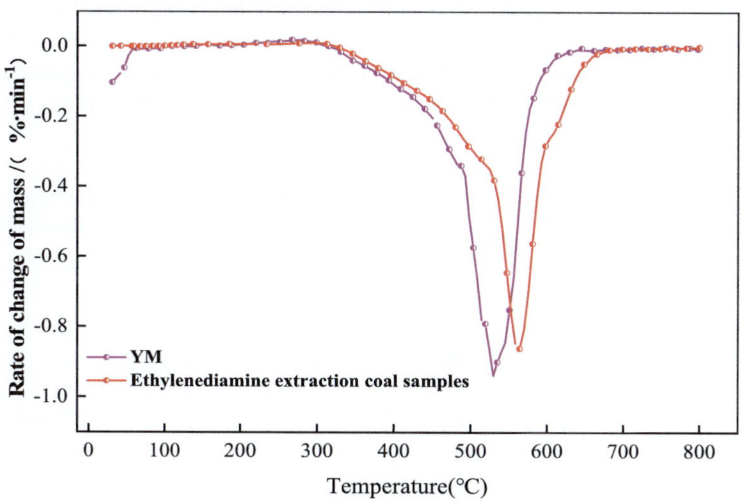

Fig. 5.6 DTG curves of YM and ethylenediamine extracted YM coal samples

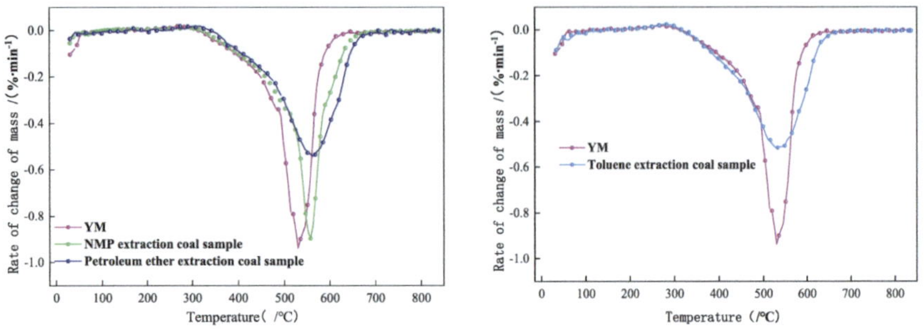

(a) YM, petroleum ether extracted YM coal samples and NMP extracted YM coal samples

(b) YM and toluene extracted YM coal samples

Fig. 5.7 Comparative analysis of DTG curves of YM coal samples

the above reactive groups is prominently lessened. This was mainly due to the fact that the number of aliphatic and oxygenated reactive groups contained in the petroleum ether-extracted and NMP-extracted coal samples was significantly reduced compared with that of the raw YM coal samples, and the above reactive groups were the key substances to promote the oxidation reaction of the samples.

Further analyzing the absolute values of the lowest peaks and valleys of the DTG curves of the YM, NMP extracted coal samples and petroleum ether extracted YM coal samples, it was found that the YM raw coal sample is larger than the NMP extracted coal sample, while the NMP extracted coal sample was larger than the petroleum ether extracted YM coal sample. This indicated that the intensity of the combustion process of the YM raw coal sample was greater than that of the NMP extracted coal sample, while the intensity of the combustion process of the NMP extracted coal sample was greater than that of the petroleum ether extracted YM coal sample. This was because the YM raw coal sample contained a greater number of reactive groups than that of the NMP extracted coal sample and the petroleum ether extracted YM coal sample, which were more prone to react with oxygen and thus react more intensely.

As can be seen from Fig. 5.7b, the DTG curves of YM and toluene-extracted YM coal samples basically coincided before 300 °C, because the oxidation reaction at this stage was mainly related to the –OH contained in the coal, and the content of –OH in toluene-extracted coal samples was almost unchanged from that in YM coal samples, so the DTG curves basically coincided at this stage. The DTG curves of YM and toluene-extracted YM coal samples were partially different between 300 and 600 °C, and it could be perceived that the absolute value of peaks and valleys of YM coal samples was larger than that of the peaks and valleys of toluene-extracted YM coal samples, which indicated that YM coal samples had reached a more intense degree than toluene-extracted YM samples, which was due to toluene having a better extraction effect on the aromatic groups in the samples, and it also extracted the –OH content of YM coal samples, and also extracted the –OH content of YM coal samples.

This is because toluene had a better extraction effect on the aromatic groups in the coal samples, and also extracted part of the aliphatic groups, so the aromatic and aliphatic groups in the toluene-extracted YM coal samples were lower than those in the YM coal samples, and the presence of aromatic groups in this stage can promote the oxidation reaction of the coal process, so the combustion reaction of the toluene-extracted YM samples was less intense than that of the YM samples in this stage, i.e., the peak and valley of toluene-extracted YM samples are smaller than that of the YM samples in terms of the absolute value of peaks and valleys. The absolute value of the peaks and valleys of the toluene extracted YM sample is smaller than that of the YM sample.

5.2.4 Comparative Analysis of TG Curves of WYM and Its Extracted Coal Samples

As can be seen from Fig. 5.9, the WYM raw coal sample and its extracted coal samples show different degrees of mass loss before 200 °C, but their mass loss is relatively small, and the mass loss of the WYM raw coal sample is larger than that of the remaining four extracted coal samples. After extraction, washing, and drying the moisture content of the four coal samples will be substantially diminished, and the active groups in the samples also appear to varying degrees of reduction, so that during the heating process Moisture volatilization and gas desorption appear to be substantially lessened compared with the WYM raw coal.

For the decomposition and combustion stage of the coal samples between 400 and 700 °C, the coal samples undergo intense oxidation reaction with oxygen, and the combustible components in the coal burn and release a large amount of gas, which leads to the expeditions mass loss of the coal samples. As can be seen from Fig. 5.8, the temperature at which the WYM raw coal reached the maximum value of mass loss rate was lower than that of the remaining four coal samples, because the WYM raw coal sample contained more active groups than the remaining four coal samples, including –OH, carbon–oxygen groups, aliphatic groups, aromatic groups, and which were contained in the coal samples and had a certain effect of promoting coal oxidation reaction, and after the extractant extracted some of the active groups from coal, the coal samples were oxidized with oxygen. When the extractant extracted some active groups from the coal, the number of active groups contained in the coal will decreased, which further affects the oxidation reaction process and the oxidation reaction rate of the coal samples, and made it necessary to reach the maximum value of the reaction rate in the combustion process of the extracted coal samples at a higher temperature.

To further investigate the effect of ethylenediamine extraction on the oxidation process of WYM, WYM and ethylenediamine extracted coal samples were compared and analyzed. As can be observed from Fig. 5.9a, from the initial temperature to 200 °C, both WYM and ethylenediamine extracted coal samples illustrated a trend

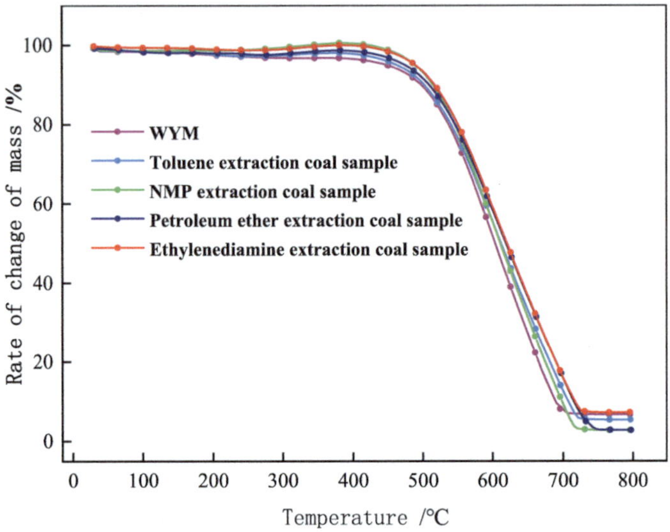

Fig. 5.8 TG curves of WYM and its extracted coal samples

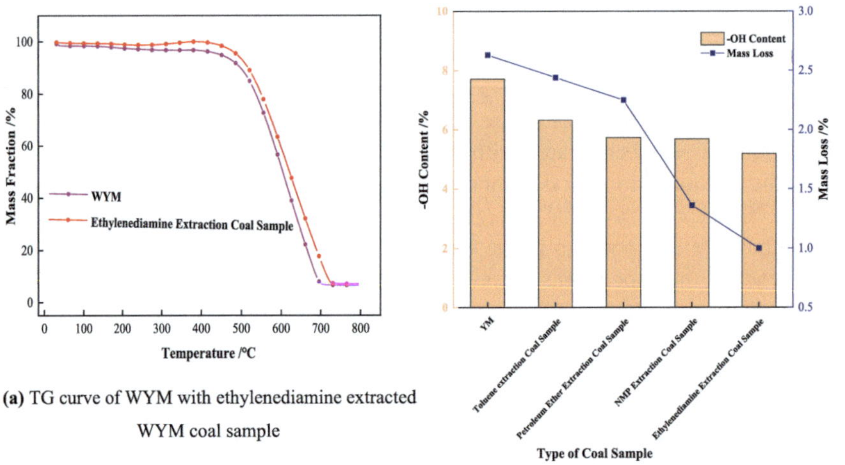

(a) TG curve of WYM with ethylenediamine extracted WYM coal sample

(b) Relationship between -OH content and mass loss of coal samples

Fig. 5.9 Comparison of TG Curves of WYM and ethylenediamine extracted WYM coal samples and the relationship between –OH content and mass loss of coal samples of mass loss, but the mass loss of the latter was saliently smaller than that of the former. This is mainly because WYM needs to be washed and dried in the process of ethylenediamine extraction, during which the impurity gas and moisture content in the coal samples will be substantially lessened. At the same time, the –OH content in the ethylenediamine-extracted coal samples decreased distinctly after the ethylenediamine extraction, so the adsorption of oxygen on the coal samples was also alleviated at this stage, and the amount of gas produced by the oxidation reaction was also reduced, which ultimately led to the loss of mass of the ethylenediamine-extracted coal samples was less than that of the WYM raw coal before 200 °C. The mass increase between 200 and 400 °C was also ascribed to the fact that the oxygen adsorption of the coal samples with the increase of temperature commenced to be larger than the gas release of the coal-oxygen composite reaction, which ultimately led to the increase of the mass of the coal-oxygen.

Between 400 and 700 °C, both WYM and ethylenediamine-extracted coal samples underwent oxidative decomposition and intense combustion reactions with further increase in temperature, resulting in a sharp decrease in quality. However, the combustion of ethylenediamine-extracted coal samples showed an obvious lag compared with that of WYM, which was mainly due to the decrease of –OH content in the coal samples after ethylenediamine extraction, which slowed down the rate of oxidation reaction of the samples, and a higher temperature was needed to provide the external energy required to achieve combustion.

Combined with the –OH content in the coal samples, the variation rules of mass loss of WYM and its extracted coal samples were further analyzed. As shown in Fig. 5.9b, there was an obvious correlation between the mass loss and –OH content of WYM and its extracted coal samples, and the higher the –OH content in the coal samples, the higher the mass loss, and vice versa, the smaller the mass loss. Compared with the raw WYM coal sample, the ethylenediamine extracted coal sample contained the smallest amount of –OH, and the toluene extracted coal sample contained the largest amount of –OH, which is a relatively active group that plays a pivoted role in the oxidation process of coal, and generates gases for desorption after the reaction to cause the loss of mass. The original WYM coal sample contained the largest amount of –OH, and therefore generated more gases for the largest loss of mass, and the raw WYM coal sample had the highest amount of –OH. The ethylenediamine extracted coal sample contained the least amount of –OH, and the mass loss becomes smaller.

The $–CH_3$ and C–O bond-containing carbon–oxygen groups in coal were active groups that play a key role in the oxidation process of coal, and it can be concluded from previous analyses that petroleum ether had a certain extraction effect on aliphatic groups, and NMP possessed a certain extraction effect on carbon–oxygen groups. Therefore, a comparison of NMP-extracted coal samples and petroleum ether-extracted samples can be used to understand the role of the $–CH_3$ and C–O bond-containing groups in the oxidation process of coal samples Fig. 5.10a.

As shown in Fig. 5.10a, the TG curves of WYM, petroleum ether-extracted coal samples and NMP ether-extracted coal samples only manifested trace changes before 200 °C, and the mass loss of WYM was the largest, while the mass loss of petroleum ether-extracted coal samples and NMP ether-extracted coal samples was smaller,

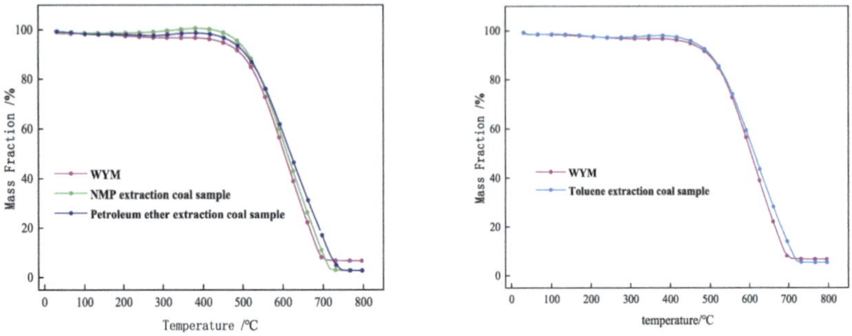

(a) WYM, petroleum ether extracted WYM coal samples and NMP extracted WYM coal samples

(b) WYM and toluene extracted WYM coal samples

Fig. 5.10 Comparative analysis of TG curves for the oxidation process of WYM coal samples

which was mainly due to the evaporation of water and gas removal at this stage, and the extraction of petroleum ether and NMP was mainly targeted at the extraction of aliphatic and carbon–oxygen groups, respectively. Petroleum ether and NMP are for aliphatic and carbon–oxygen groups, respectively, and the main purpose of this stage is to promote the reaction between –OH and oxygen to generate gas desorption, which leads to the smallest mass loss of NMP ether-extracted coal samples. Between 200 and 400 °C, the mass gain of coal samples was also due to the adsorption of oxygen by the reactive groups in coal. In the decomposition and combustion stage of the coal samples between 400 and 700 °C, the combustion temperatures of the petroleum ether extracted coal samples and the NMP ether extracted coal samples were lagged, which was mainly due to the partial reduction of the –OH content in the coal samples after the petroleum ether and the NMP extraction, and the reduction of the –OH led to the lag in the combustion temperatures of the petroleum ether extracted coal samples and the NMP ether extracted coal samples, and meanwhile, the aliphatic groups in the coal samples were attenuated by the petroleum ether extraction, which led to the lag of the combustion temperatures of the petroleum ether extracted coal samples. At the same time, after petroleum ether extraction, the aliphatic groups in the coal samples were curtailed, which further led to the lag in the combustion temperature of the petroleum ether extracted coal samples.

From Fig. 5.10b, it can be seen that the curve changes of WYM and toluene extracted coal samples are basically the same before 200 °C. This was because the content of aromatic and aliphatic groups in toluene extracted coal samples was lessened compared with that of WYM coal samples, and the most critical group that affects the oxygen adsorption and gas desorption of the samples before 200 °C is –OH, and the content of –OH in toluene extracted coal samples had almost no change compared with that of the WYM raw coal samples. There was almost no change in the –OH content of the toluene extracted coal sample compared with that of the WYM raw coal sample, so the curve changes of the toluene extracted coal sample and the WYM raw coal sample had no obvious change before 200 °C. Between 200 and 400 °C, the mass gain trend of WYM and toluene extracted coal samples was also due to the adsorption of oxygen. Between 400 and 750 °C, the decomposition and combustion reaction of WYM and toluene extracted coal samples occurred. In this reaction stage, the aromatic groups in the coal affect the oxidation rate of the coal samples, and since toluene had a certain extraction effect on the aromatic groups, the maximum temperature of the combustion reaction rate of the toluene-extracted coal samples corresponded to a higher temperature compared with that of the WYM, and the mass loss curves of the toluene-extracted coal samples between 600 and 750 °C were shifted to the high temperature region.

5.2.5 Comparative Analysis of DTG Curves of WYM and Its Extracted Coal Samples

Similarly, the TG curves of WYM and extracted coal samples were differentiated to the first order to obtain the DTG curves of the coal samples, and the trends of the DTG curves could be used to further analyze the mass change of oxidation reaction and the reaction rate of different coal samples during the heating process.

Similar to the YM and ethylenediamine-extracted coal samples, after ethylenediamine extraction, the temperature corresponding to the maximum mass loss rate of the ethylenediamine-extracted coal samples showed a notable increase compared with that of the WYM raw coal, suggesting that the ethylenediamine-extracted coal samples were more difficult to undergo oxidation reactions than the WYM. In addition, the absolute values of peaks and valleys around 600 °C were larger for WYM than for the ethylenediamine-extracted coal sample, indicating that the combustion process of the WYM coal sample was more intense than that of the ethylenediamine-extracted coal sample. This was due to the fact that the reaction rate in the ethylenediamine residual sample was slowed down due to the lower content of –OH, and the intensity of the reaction was reduced.

The effects of aliphatic groups and carbon–oxygen groups on the oxidation reaction rate of coal samples were further analyzed by comparing the DTG curves of WYM, petroleum ether-extracted WYM coal samples, and NMP-extracted WYM coal samples in Fig. 5.11. The peaks and valleys of the DTG curves of the three coal samples in Fig. 5.12a correspond to the temperatures at which the fastest oxidation reaction rate occurs in the coal samples, and the values of the peaks and valleys represent the magnitude of the reaction rate. After petroleum ether and NMP extraction, the oxidation reaction rates of petroleum ether-extracted WYM coal samples and NMP-extracted WYM coal samples decreased, with the petroleum ether-extracted coal samples decreasing the most.

As can be seen from Fig. 5.12b, the DTG curves of WYM and toluene-extracted WYM coal samples are basically the same before 200 °C. This is because the reaction rate of WYM coal samples is mainly affected by the –OH contained in the coal in this stage, while toluene has some extraction effect on the aromatic groups of the

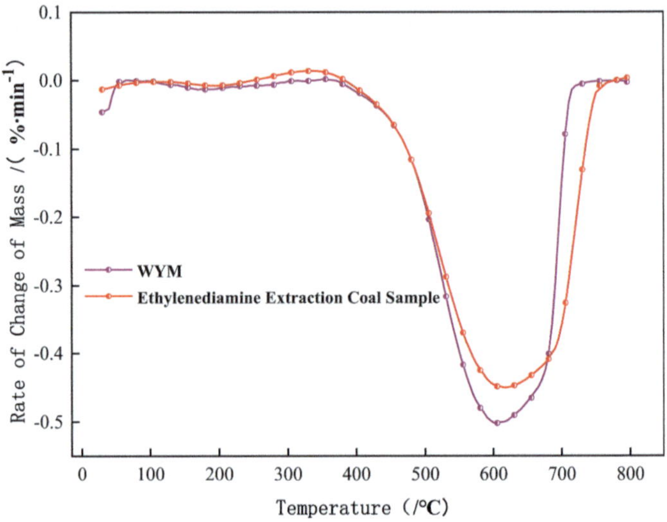

Fig. 5.11 DTG curves of WYM and ethylenediamine extracted WYM coal samples

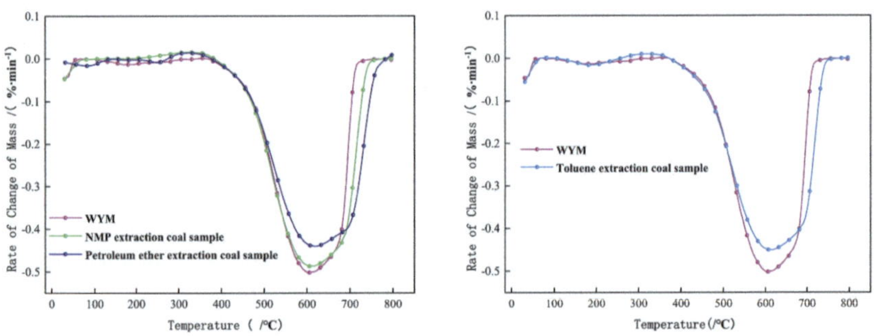

(a) WYM, NMP extracted WYM coal samples and petroleum ether extracted WYM coal samples

(b) WYM and toluene extracted WYM coal samples

Fig. 5.12 Comparative analysis of DTG curves of WYM coal samples

coal samples, and has a poorer extraction effect on the –OH in the coal samples, so the reaction trend of the coal samples is similar to that of the raw coal of WYM in this stage. The reaction trend in this stage is similar to that of WYM raw coal. Between 400 and 700 °C, the temperatures corresponding to the lowest peaks and valleys of the DTG curves, the toluene-extracted coal samples were skewed to higher temperatures compared with WYM, and the absolute values of the very exceedingly small points corresponding to the peaks and valleys were significantly lower, which indicated that the aromatic groups had a substantial effect on the combustion process and the combustion reaction rate of the coal samples.

5.3 Characteristics of Thermal Effect Changes Before and After Extraction of Coal Samples

To further understand the changes of the heat absorption and excretion parameters of the coal samples during the oxidation process before and after extraction, the thermal effects of the raw coal and its extracted coal samples during the oxidation process were analyzed by differential scanning calorimetry (DSC). The mass of the coal samples was converted to the same scale by taking the original coal as the base and combining the extraction rates of different extractants on the coal samples in Chap. 3. In the DSC curves, if the coal samples appeared exothermic, the curves showed upward peaks, and the heat absorption showed downward peaks. Integrating the exothermic peaks in the DSC curve, the exothermic or absorptive heat of the coal sample at that stage can be obtained.

5.3.1 Definition of Coal Enthalpy

In the testing process, by measuring the temperature difference and heat flow difference between the coal sample and the control and converting the information into graphs, i.e., DSC curves, the heat generated per unit mass of the coal sample during the oxidation process, i.e., enthalpy, can be acquired by integrating the peak areas of different reaction stages on the DSC curves.

Figure 5.13 exhibits the DSC curves of two coal samples, YM and WYM, under the conditions of air atmosphere and 10 °C/min temperature rise. Comparative analyses of the DSC curves can receive the differences in the amount of heat released at different stages of the oxidation process due to the different degrees of metamorphism.

As can be seen from Fig. 5.13, the DSC curves of YM and WYM are basically the same between 30 and 200 °C, and the DSC curves of YM and WYM were obviously different after 200 °C, and YM commenced to manifest obvious exotherm after 200 °C, and it reaches the peak of the first exotherm at roughly 370 °C, and the peak of the second exotherm at about 505 °C, which are caused by thermal decomposition and combustion, respectively, and due to combustion. This was caused by the thermal decomposition and combustion of YM, respectively, and the area of the second exothermic peak was much larger than that of the first one because the combustion reaction was more intense. Compared with YM, the DSC curves of WYM showed obvious hysteresis phenomenon, and the peak shapes are also quite different, the first exothermic peak of WYM occurred at ~450 °C, and the peak shape was smaller, and the second exothermic peak occurred at ~605 °C, and the peak shape was broader. This is mainly because WYM has a higher degree of metamorphism, with an increased proportion of carbon and a decreased proportion of oxygen-containing reactive groups, and thus required higher temperatures for decomposition and combustion reactions to occur, resulting in a backward shift of

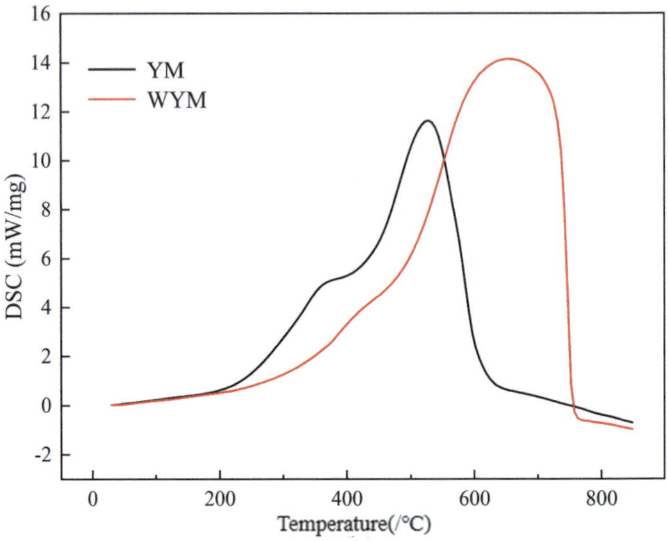

Fig. 5.13 DSC curves for YM and WYM

the exothermic peaks. In addition, in the combustion stage, the combustion lasted longer and releases more heat due to the higher carbon content of WYM, which resulted in a broader peak shape and larger peak area of the second exothermic peak, and the overall exothermic amount was higher than the exothermic amount of YM.

5.3.2 *Comparative Analysis of DSC Curves of YM and Its Extracted Coal Samples*

In the oxidation process of YM and its extracted coal samples, the absorbed and exothermic parameters of the coal samples as the key parameters reflect the influence of different extractants on the exothermic amount of YM coal samples in the oxidation process. The DSC curves of YM and its extracted coal samples were comparatively analyzed, and the results are shown in Fig. 5.14.

By comparing and analyzing the changes in the DSC curves of the original YM coal sample and the extracted residual coal, it can be seen that the changes in the content of each group in the coal samples have different influences on the heat absorption and exotherm of the coal samples in the oxidation process. As can be seen from Fig. 5.14, the exothermic trend of the YM original coal sample and the extracted residual coal sample during the heating process was basically the same, and both of them started to appear exothermic trend at ca. 200 °C, and the exothermic of the YM original coal sample reached the first peak at 370 °C, and then illustrated a stable exothermic state between 370 and 420 °C, and then the DSC curves of the

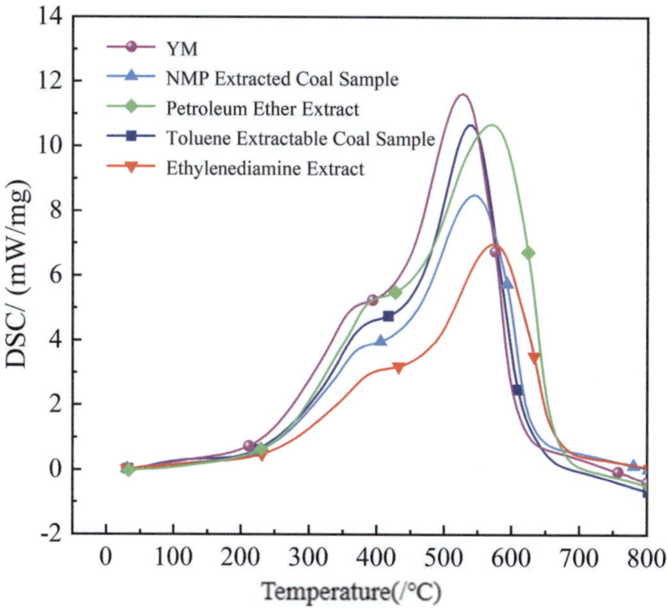

Fig. 5.14 DSC curves of YM and its extracted coal samples

samples commenced to rise sharply after 420 °C, and reached the peak at about 527 °C. The DSC curve started to increase sharply after 420 °C and reached the peak at about 527 °C. However, the exothermic amount of YM coal samples during the heating process after extraction showed obvious changes, and the effects of different extractants on the exothermic amount of coal samples also differed greatly. To further analyze the effects of different extractants on the exothermic amount of YM, we compared the DSC of different extracted coal samples with the original YM coal.

Figure 5.15 shows the comparison of the DSC of YM coal and ethylenediamine residual coal samples. It can be seen from the figure that in the extire heating and oxidation process, the DSC value of ethylenediamine residual coal samples is smaller than that of YM coal samples, and the DSC value of ethylenediamine residual coal samples is slightly smaller than that of YM coal samples within the range of 30–200 °C, and there are two types of reactions, namely, the heat absorption reaction of water evaporation and desorption and the coal-oxygen composite exothermic reaction. Therefore, the DSC value in this stage was the value after subtracting the heat absorption from the heat release. The decrease of DSC value is due to the fact that the ethylenediamine extracted coal samples after ultrasonic extraction treatment, ethylenediamine extracted the hydroxyl group in the coal and slowed down the coal-oxygen complex exothermic reaction, which caused the DSC curve of the ethylenediamine extracted coal samples to be lower than that of the original coal samples of YM in this stage. After 200 °C, the water evaporation and desorption reaction of coal basically ended, and the reaction of active group and oxygen complex mainly appeared in coal, with the increase of temperature, the activity of active group and oxygen complex reaction gradually increased, and the intensity of heat release began to increase, and the DSC curve of the YM coal sample reached the first peak at roughly 370 °C, due to the rate of reactive group and oxygen complex reaction in the structure of the YM coal sample reached the maximum, and the DSC curve of the YM coal sample reached the first peak at ~370 °C, and the DSC curve of the YM coal sample was lower than that of the original coal sample. This temperature also corresponds to the temperature point of the mass maximum of the coal sample on the TG curve.

As can be seen from the figure, the DSC value of the ethylenediamine-extracted coal sample at this point was obviously smaller than that of the YM original coal sample, which was because ethylenediamine extracts –OH from the coal, but the hydroxyl radicals were able to react with the other reactive sites in the coal during the process of heating and oxidation reaction, and capture the H of the reactive sites to emit heat, thus leading to the reduction of the ethylenediamine extracted coal sample due to the reduction of the –OH content, and thus the degree of the oxidation reaction was not as intense as the YM original coal at this stage. The degree of oxidation reaction in this stage is not as intense as that of YM raw coal, so the exothermic amount is reduced compared with that of YM raw coal. The exothermic value of YM reaches the peak at about 527 °C, which also corresponds to the peak and valley of the fastest mass loss in the DTG curve, and it can be seen that the DSC peak of the ethylenediamine extracted coal samples decreased significantly compared with that of the original coal of YM, and the decrease in the DSC peak

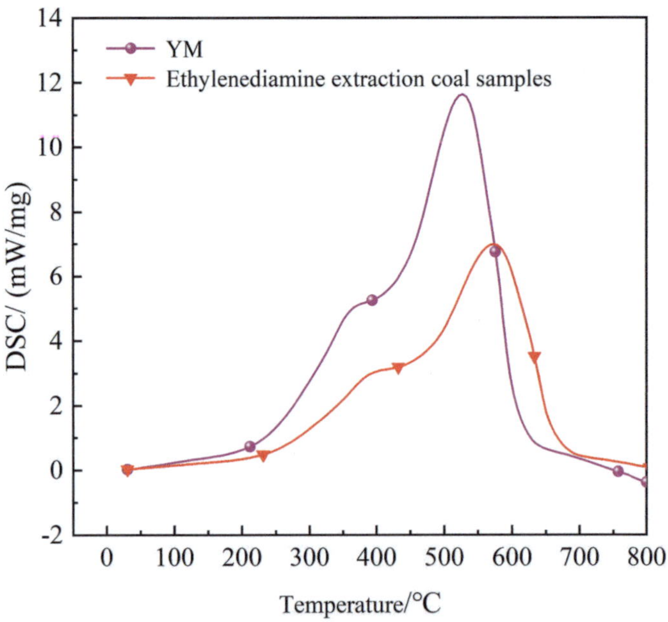

Fig. 5.15 DSC curves of YM and ethylenediamine extracted YM coal samples

is mainly due to the fact that the reactive –OH groups in the coal still played a key role in the combustion stage, so the reduction of the reactive –OH groups led to the decrease in the DSC peak of the ethylenediamine extracted coal samples at the stage of oxidation. The combustion reaction of ethylenediamine extracted coal samples was not vigorous enough and the reaction rate was slow, which was also consistent with the TG curve. Ultimately, the DSC of the whole ethylenediamine extracted coal sample was shifted to a higher temperature compared with YM, and the exothermic amount was pronouncedly reduced.

The heat release power of coal samples during the oxidation reaction can be obtained by further integrating the DSC curves of different coal samples (Fig. 5.16).

In the preliminary stage of the oxidation of YM coal samples (before 200 °C), the main reactions occurring in the coal samples are heat volatilization of moisture, desorption of adsorbed gases in the coal pores, and a minute amount of coal-oxygen composite reaction, in which the volatilization of moisture and the desorption of gases are both heat-absorbing reactions, while the coal-oxygen composite was an exothermic reaction. From the DSC curves of different samples, it can be perceived that the intensity of the exothermic reaction was greater than the intensity of the heat-absorbing reaction in the early stage of the reaction, so the coal samples as a whole showed an exothermic process. With the gradual increase of temperature, the intensity of water evaporation and desorption reaction of coal samples gradually decreased, and the intensity of coal-oxygen composite reaction gradually strengthened, and a large number of active groups in the coal were activated to compound with oxygen to release heat, and the temperature at this time was the initial exothermic temperature of the coal samples when the coal-oxygen composite reaction releases the same heat as that of evaporation and desorption reaction. With the further increase of temperature, in the interval of 200–700 °C, the coal samples mainly experienced the coal-oxygen compounding stage, the thermal decomposition stage and the combustion stage, and the reactions in these three stages are all exothermic reactions. Among these three stages, the chemical reaction in the combustion stage is the most intense and releases the most heat, so the highest point on the DSC curve was the point where the intensity of the combustion reaction of the coal sample is the largest, and the horizontal coordinate of this point is the maximum exothermic temperature, and the vertical coordinate is the maximum heat-releasing power, and the sum of the total exothermic amount of these three stages is the exothermic amount shown in the figure. The specific DSC parameters of the samples are shown in the following table (Table 5.3).

To observe the difference between the maximum exothermic temperatures of YM and its extracted coal samples more intuitively, the data were plotted as a histogram (Fig. 5.17). As can be seen from the figure, the maximum exothermic temperature of the original YM coal sample was 527.54 °C, and the maximum exothermic temperatures of the coal samples treated with different extractants were all larger than that of the original YM coal sample, the ethylenediamine-extracted coal sample showed the largest growth, with the maximum exothermic temperature of 573.95 °C, the petroleum ether-extracted coal sample showed the second largest growth, with the maximum exothermic temperature of 571.63 °C, and the toluene-extracted coal

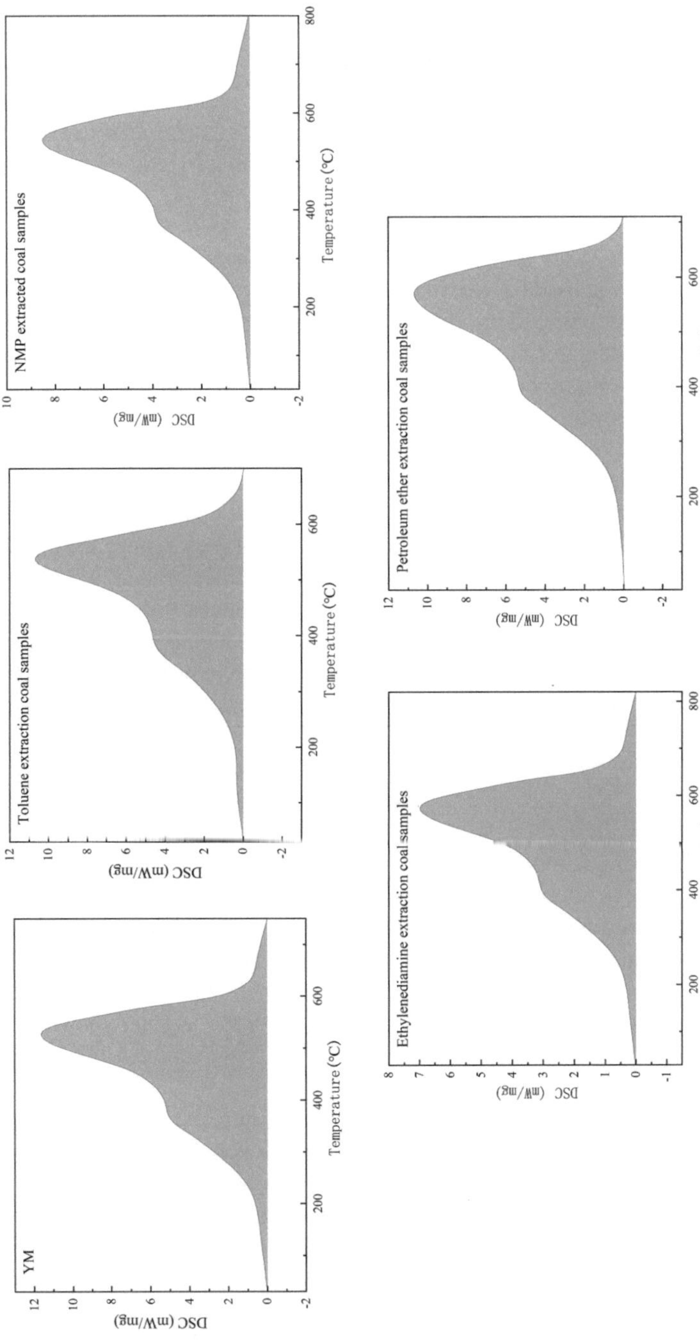

Fig. 5.16 DSC parameter profiles of YM and its extracted coal samples

Table 5.3 DSC parameters of YM and its extracted coal samples

Samples	Initial exothermic temperature (°C)	Maximum heat release power (mW/mg)	Maximum exothermic temperature (°C)	enthalpy (J/g)
YM	137.48	11.63	527.54	2330.62
Toluene extraction YM coal sample	162.76	10.66	535.63	2016.25
Petroleum ether extraction YM coal sample	132.86	10.68	571.63	2154.74
Ethylenediamine extraction YM coal sample	141.17	6.98	573.95	1552.48
NMP extraction YM coal sample	145.42	8.50	543.24	1847.12

sample had the smallest growth, with the maximum exothermic temperature of 535.63 °C, and the largest exothermic temperature of 535.63 °C. The maximum exothermic temperature was 535.63 °C, and the maximum increase was ca. 8%. The coal samples in the combustion process reaction speed was exceptionally fast, the exothermic amount was also larger, the temperature change of the coal samples is also faster, so the difference between the maximum exothermic temperature of the extracted coal samples and the original coal samples in this range can be regarded as the extractant treatment had no effect on the maximum exothermic temperature of the coal samples. This was due to the fact that the extractant treatment mainly extracts the small molecule reactive groups in the structure of the coal sample, while the oxidation of the large molecule skeleton structure of the coal mainly released a large amount of heat at the maximum exothermic temperature. Therefore, for the same type of coal samples, under the same oxidation conditions, the temperature at which the macromolecular skeleton of the coal samples participates in the combustion reaction was certain, and the small molecular reactive groups had little influence on the maximum exothermic temperature of the coal samples.

The maximum exothermic power of YM and its extracted coal samples was further analyzed, and the difference between the maximum exothermic power and the maximum exothermic power of YM and its extracted coal samples can be seen in Fig. 5.18, which shows that the maximum exothermic power of YM is 11.63 mW/mg, and the extracted coal samples have a decreasing tendency compared with that of the original samples, and the maximum exothermic power of four samples, namely, toluene, petroleum ether, NMP, and ethylenediamine is 10.66, 10.68, 8.5, 6.5 and 6.6 mW/mg, respectively. This was because the maximum heat release power of coal samples at this temperature was composed of the oxidative exotherm of the macromolecular skeleton and the oxidative exotherm of the active group, so the content of the active group in the YM coal samples decreased to different degrees after extraction, which led to the decrease in the heat release power. The best extraction effect

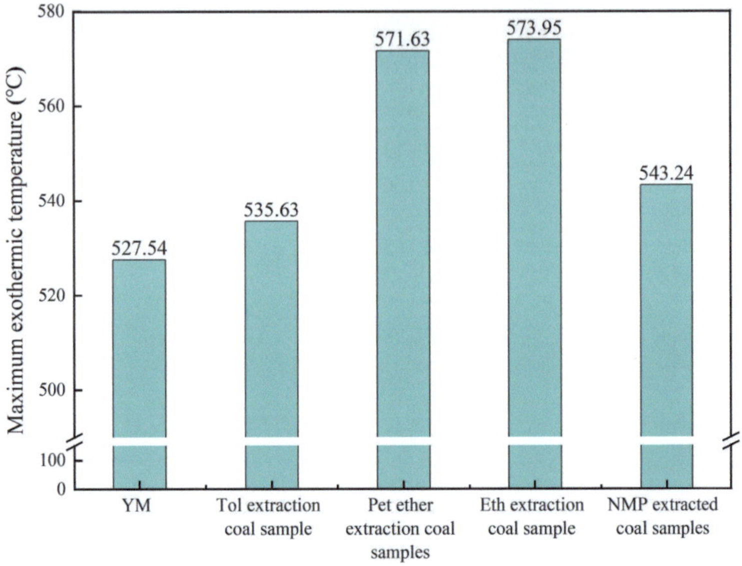

Fig. 5.17 Maximum exothermic temperature diagram for YM and its extracted coal samples

of ethylenediamine on –OH led to the most obvious decrease of –OH content in ethylenediamine-extracted coal samples, and thus the most obvious decrease of its maximum heat release power, while toluene and petroleum ether had the least effect on the maximum heat release power of coal samples because they mainly targeted the more difficult to extract aromatic and aliphatic groups and had poor extraction effect.

Figure 5.19 compares and analyses the exothermic data of YM and its extracted coal samples. As can be observed from the figure, the exergy of YM original coal sample was 2330.62 J/g, and the exergy of all the extracted coal samples exhibited different degrees of decrease, the petroleum ether extracted coal sample showed the smallest decrease, which was ca. 175.88 J/g; the exergy of ethylenediamine extracted coal sample showed the largest decrease, which was 778.14 J/g; and the decreases of NMP and toluene extracted coal samples were in the middle of the range. The decrease of NMP and toluene extracted coal samples is in the middle. The main reason was that a large number of key reactive groups (e.g. –OH, carbon–oxygen reactive groups, aliphatic groups, and aromatic groups), which can react with oxygen in a complex and exothermic manner, were dissolved and extracted out by the extractant, which led to a decrease in the number of reactive groups participating in the reaction during the holistic oxidation process of the coal samples, and a decrease in the total exothermic amount of the coal samples in the oxidation process. In addition, the extraction rate of active groups varied among different extractants, resulting in different reduction of active groups in YM, and thus the net exothermic capacity of the extracted coal samples decreased to different degrees.

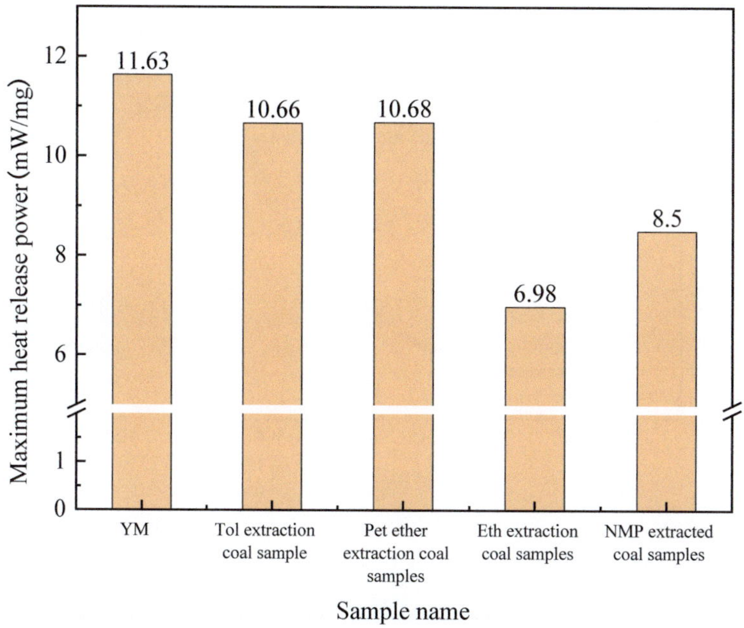

Fig. 5.18 Maximum heat release power of YM and its extracted coal samples

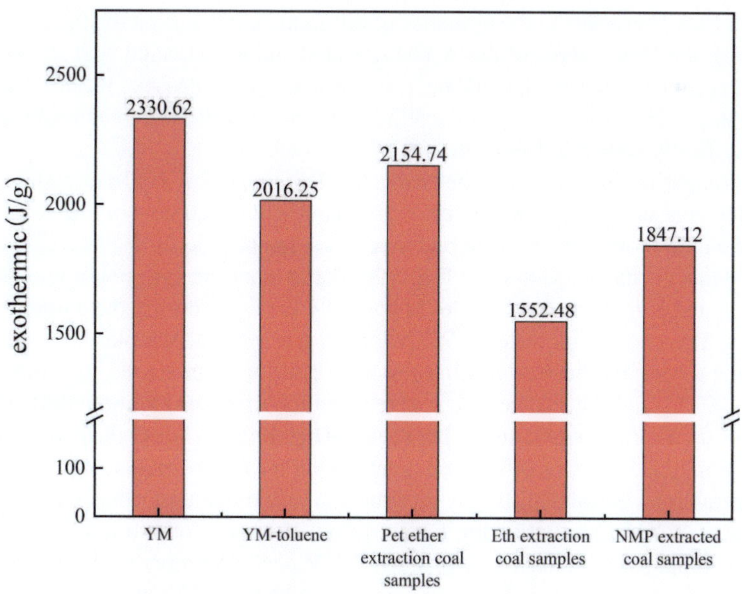

Fig. 5.19 Exergy diagram of YM and its extracted coal samples

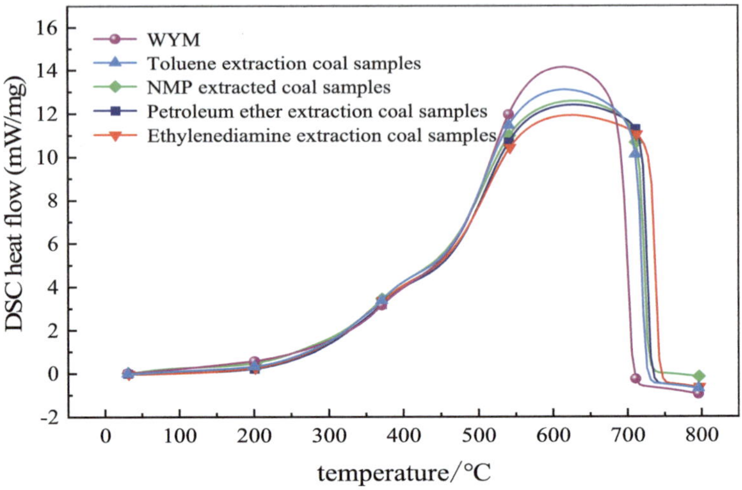

Fig. 5.20 Summary of DSC curves for WYM and its extracted coal samples

5.3.3 Comparative Analysis of DSC Curves of WYM and Its Extracted Coal Samples

In order to further analyze the influence of different extractants on the thermal effect of WYM, the DSC curves of the WYM raw coal and its extracted coal samples are integrated, and we can get the total heat absorption or heat release of the coal samples at this stage. The DSC curves of the WYM raw coal and its extracted coal samples are specifically shown in Fig. 5.20.

By comparing and analyzing the changes of DSC curves of WYM raw coal and its extracted coal samples, the influence of the changes of the content of different active groups on the exothermic absorption of the oxidation process of WYM coal samples can be seen. As can be seen from Fig. 5.20, the exothermic variation rule of WYM raw coal and extracted coal samples is basically the same during the entire heating process. The exothermic rule of WYM and extracted coal samples was basically the same during the holistic heating process. Both of them started to exotherm at about 200 °C, and the exothermic rule reached the first peak at about 380 °C, and the exothermic rule was reduced between 380 and 450 °C, and then rose slowly. In addition, the effects of different extractants on the exothermic amount of WYM after extraction were also distinctly different, and the maximum value of exothermic amount of WYM after extraction was reduced, and the exothermic peaks became broader compared with that of the original WYM coal. To further analyze the effects of different extractants on the exothermic amount of WYM in detail, they were compared and analyzed separately.

Figure 5.21 shows the comparison of DSC curves between WYM and ethylenediamine extracted WYM coal samples. On the whole, the DSC values of the ethylenediamine extracted coal samples are lower than those of the original WYM coal samples. Especially in the temperature range of 100 and 300 °C, the DSC values of the ethylenediamine extracted coal samples were conspicuously smaller than those of the original YM coal samples. In this temperature range, the moisture volatilization and heat absorption reaction and the coal-oxygen complex exothermic reaction occurred simultaneously in the coal samples. Since these two reactions existed at the same time, the DSC value at this stage was the difference between the heat absorption and the heat release. After ultrasonic extraction, ethylenediamine was able to extract –OH from WYM coal, which slowed down the coal-oxygen complex exothermic reaction to a certain extent, resulting in the DSC curves of ethylenediamine-extracted coal samples being lower than those of the original coal samples of YM in this stage. When the temperature exceeded 300 °C, the main reaction in the coal was the compound reaction between active groups and oxygen, and with the continuous increased of temperature, the activity of this compound reaction gradually increased, and the intensity of heat release began to increase. When the temperature reached 415 °C, a peak appears in the DSC curve of WYM coal samples, which is due to the fact that the rate of the complex reaction between the active groups and oxygen in the structure of WYM coal samples reaches the maximum. This temperature point also corresponded to the temperature point of the coal sample mass maximum on the TG curve. The exotherm of the WYM samples reached a maximum at ~615 °C, which corresponded to the maximum mass loss rate on the DTG curve.

Comparative analysis of the DSC curves of the ethylenediamine-extracted WYM coal samples reveals that the DSC values of the ethylenediamine-extracted WYM coal samples were obviously smaller than those of the original WYM coal, which was mainly due to the fact that after the extraction of ethylenediamine, the –OH content of the WYM decreases, and –OH is the key reactive group to promote the oxidation of the multi-carbon substances in the coal, so that the heat-induced oxidation of

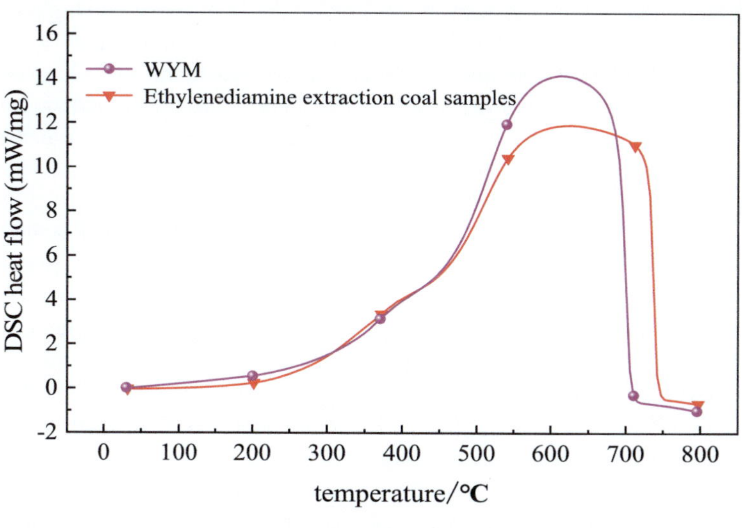

Fig. 5.21 Comparative DSC analysis of WYM and ethylenediamine extracted WYM coal samples

ethylenediamine-extracted WYM samples has been suppressed to a certain degree, leading to a reduction of its reaction rate, and the maximum release rate was lessened. Therefore, the thermal oxidation process of ethylenediamine extracted WYM coal samples was inhibited to a certain extent, which led to the reduction of its reaction rate and the maximum exothermic power. It is worth noting that the lack of -OH in the ethylenediamine-extracted WYM coal samples requires higher heat to initiate the combustion of the coal macromolecular skeleton structure during the oxidation process, so the exothermic curves are shifted to the high temperature region. The exergy of the coal samples during the oxidation reaction can be obtained by further integrating the DSC curves of different coal samples Fig. 5.22.

In the early stage of coal oxidation (i.e., before 200 °C), the coal samples mainly undergo moisture evaporation, desorption of adsorbed gases in the coal pores, and a trace amount of coal-oxygen complexation. Among them, moisture volatilization and gas desorption are adsorptive reactions, while coal-oxygen complexation is an exothermic reaction. From the DSC curves of WYM and its extracted coal samples, it can be observed that at this stage, the intensity of the heat-absorbing reaction was smaller than that of the exothermic reaction, so the coal samples as a whole were exothermic. With the gradual increase in temperature, the intensity of the moisture volatilization and desorption reaction of the coal samples gradually decreased, while the coal-oxygen complex reaction was gradually enhanced.

When a large number of active groups in the coal were activated and compounded with oxygen to release heat, and this exothermic amount was comparable with the heat absorption of evaporation desorption, the temperature was the initial exothermic temperature of the coal sample. When the temperature was further increased to the range of 200–630 °C, the coal sample mainly undergoes three stages of coal-oxygen compounding, thermal decomposition and combustion, which were all exothermic reactions. Among them, the chemical reaction in the combustion stage was the most intense and releases the most heat. Therefore, on the DSC curve, the highest point represented the moment when the intensity of the combustion reaction of the coal sample was the greatest, the horizontal coordinate of this point was the maximum exothermic temperature, and the vertical coordinate reflected the maximum heat release power. The sum of the total heat release of these three stages was the heat release shown in Fig. 5.22. Table 5.4 summarizes the specific DSC parameters for WYM and its extracted coal samples.

As we can be seen in Fig. 5.23, the temperature of the WYM original coal sample was 615.45 °C when it reached the maximum exothermic temperature. In comparison, the maximum exothermic temperatures of the coal samples extracted by the extractant, except for the toluene extracted WYM coal sample, were higher than that of the WYM original coal sample. Among them, the maximum exothermic temperature of the coal samples after ethylenediamine extraction increased the most, which was ca. 627.05 °C. The maximum exothermic temperatures of the petroleum ether and NMP extracted coal samples were about the same, and the maximum exothermic temperature was around 625 °C. Given the quick reaction rate, high exotherm and rapid temperature change of the coal samples during combustion, the temperature difference in this range can be considered as a negligible effect of the extraction

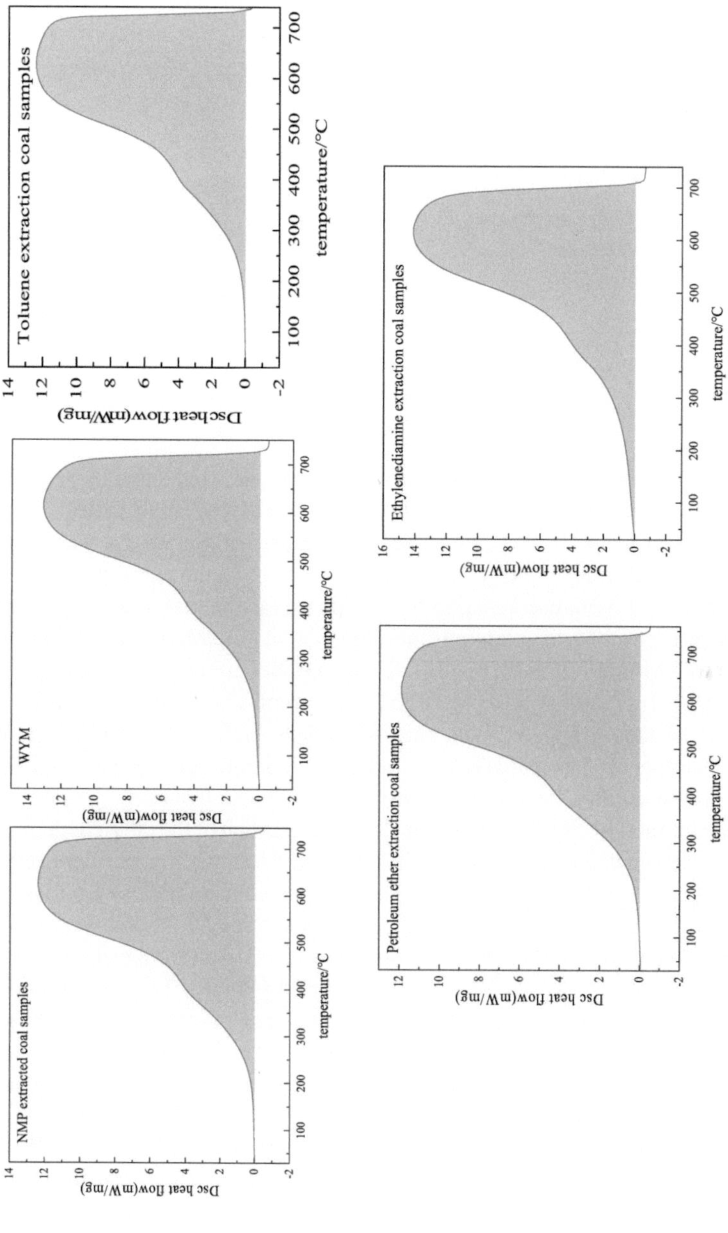

Fig. 5.22 Curves of DSC parameters for different samples

Table 5.4 DSC parameters of WYM and its extracted coal samples

Specimens	Initial exothermic temperature (°C)	Maximum heat release power (mW/mg)	Maximum exothermic temperature (°C)	Enthalpy (J/g)
WYM	97.92	14.15	615.45	3554.79
Ethylenediamine extraction WYM coal samples	102.42	11.91	627.05	3476.5
NMP extraction WYM coal sample	108.16	12.56	626.05	3444.11
Petroleum ether extraction WYM coal sample	92.11	12.39	625.48	3489.34
Toluene extraction WYM coal sample	110.74	13.09	615.38	3444.26

treatment on the maximum exothermic temperature of the coal samples. This was mainly due to the high degree of metamorphism of WYM, the small molecule reactive group content in the coal was minute, and the extraction process was mainly targeted at the small molecule reactive group in the coal, so the extraction has little effect on the overall oxidation reaction of WYM, and as the maximum exothermic temperature was reached, it is mainly the combustion of large molecule skeleton structure of WYM that released the heat. Therefore, the effect of extraction on the maximum exothermic temperature of WYM coal samples was not noteworthy.

As we can be seen in Fig. 5.24, the maximum heat release power of the original WYM coal sample was 14.15 mW/mg, and the maximum heat release power of the extracted coal samples decreased slightly compared with that of the original coal samples after extraction, and the maximum decrease was that of the ethylenediamine-extracted residual coal, with a decrease of 2.24 mW/mg, followed by petroleum ether- and NMP-extracted samples, with a decrease of the same magnitude of 1.6 mW/mg, while the maximum heat release power of the toluene-extracted coal samples lessened the least with 1.06 mW/mg. This was also similar to the reason for the change of the maximum heat release power of YM and its extracted samples. The lessened of the maximum heat release power after toluene extraction was the smallest, roughly 1.06 mW/mg, which was similar to that of the YM and its extracted coal samples, due to the fact that the maximum heat release power of the coal samples was contributed by two parts, the oxidative exotherm of the macromolecular skeleton structure of the coal, and the oxidative exotherm of the reactive groups. When the WYM coal samples were extracted, the active groups were dissolved by the extractant, which led to a decrease in the oxidative exotherm of the active groups in the coal samples, and the maximum heat release power decreased compared with that of the original coal samples. However, owing the difference in the extraction effect of different extractants on the active groups, the decrease in the maximum heat release power

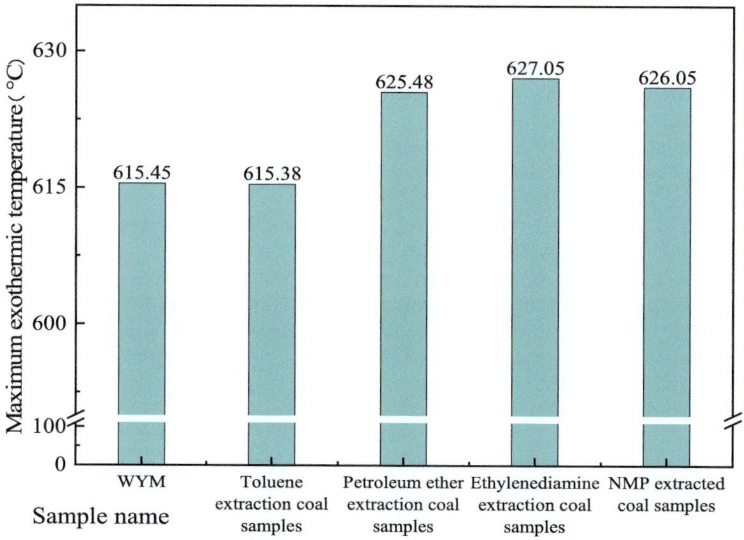

Fig. 5.23 Maximum exothermic temperature plots for WYM and its extracted coal samples

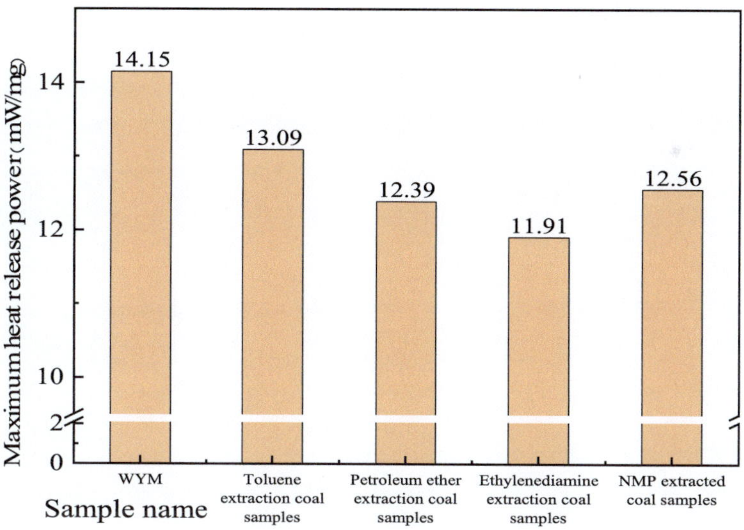

Fig. 5.24 Maximum heat release power plot for WYM and its extracted coal samples

of the coal samples was also different, the better the extraction effect, the larger the decrease in the maximum heat release power; the worse the extraction effect, the smaller the decrease. Therefore, as a whole, owing to the absence of some active groups in the extracted coal samples, the overall exothermic range became broader and the maximum exothermic power was abated.

References

Bai G, Zhou XH, Song DH et al (2017) Research on coal's combustion characteristics and kinetics parameters as a function of its metamorphic degree. China Saf Sci J 27(09):63–68

Deng J, Li Q, Shu C et al (2017) Experimental study on the thermal properties of coal during pyrolysis, oxidation, and re-oxidation. Appl Therm Eng 110:1137–1152. https://doi.org/10.1016/j.applthermaleng.2016.09.009

Hao CS, Yuan YC, Jia TG et al (2022) Infrared spectral research on chemical structure of coal with different levels of metamorphism. Saf Coal Mines 53(11):15–22

Jia TG, Zhang ZZ, Hao CS (2024) Study on spontaneous combustion characteristics of bituminous coal with different degrees of metamorphism. Saf Coal Min 55(02):87–98

Shu P, Zhang Y, Deng J et al (2023) Study on the effect of mixed particle size on coal heat release characteristics. Int J Coal Prep Utiliz 12:8810. https://doi.org/10.1080/19392699.2022.2078810

Song J, Deng J, Zhao J (2021) Comparative analysis of exothermic behaviour of fresh and weathered coal during low-temperature oxidation. https://doi.org/10.1016/j.fuel.2020.119942

Wang W (2005) Application of the thermogravimetric analysis in the research of the oxidation procession of coal. Xi'an University of Science and Technology

Wang K (2015) Study on the oxidation reaction and kinetics at low temperature of Jurassic coal in north Shaanxi. Xi'an University of Science and Technology

Yang Y, Fei J, Luo Z (2023) Experimental study on characteristic temperature of coal spontaneous combustion. J Therm Anal Calorim 148:10011–10019. https://doi.org/10.1007/s10973-023-12365-0

Chapter 6
Molecular Dynamics Characterization of the Evolution of Reactive Groups in Coal Oxidation Reactions

Abstract By combing molecular dynamics simulation with experiments, oxidation reacting pathways of active groups as well as its Molecular dynamics characteristics were investigated comprehensively. A new coal spontaneous combustion theory titled "Chain Self-promoted Oxidizing Coal Spontaneous Combustion Theory Induced by Active Groups" was proposed. This proposed theory interpreted the generative mechanisms of index gases were explained, and correlated reaction pathways of active groups to characteristics of coal spontaneous combustion. This chapter not only completes the current theory of coal spontaneous combustion but also sets a scientific guidance for developing new prevention and control techniques of coal spontaneous combustion.

The aim of this chapter is to probe the evolution of the reactive structure in the coal oxidation reaction, and to determine the active sites of the reactive groups of the coal oxidation reaction, and to analyze the transition states of the reactants and the reaction pathways, and then to correlate the coal auto-ignition reactive groups with the indicator gases by testing the parameters of the gas products in the process of the coal oxidation reaction through TG-MS experiments, and to further investigate the oxidation reaction properties of the coal auto-ignition reactive groups and indicator gases. In this chapter, based on the molecular structure model of coal reactive groups constructed in Chap. 3, the oxidation mechanisms of the crucial reactive groups (–OH, –CHO, $–CH_2CH_3$) in the YM structure and the important reactive groups (–OH, –CHO, $–CH_3$) in the WYM structure were investigated. Gaussian 16 software was adopted to optimize the geometry of the model and observe and analyze its molecular orbital properties and electron cloud distribution, to clarify the reactivity of the reactive groups, to analyze the alternative of bond energy and bond angle during the reaction process, and to simulate and calculate the thermodynamic parameters such as apparent activation energy and enthalpy change of the reactants, so as to summarize the oxidation reaction mechanism of the reactive groups in the process of coal oxidation.

J. Deng et al., *Coal Spontaneous Combustion Theory with Chain Self-promoted Oxidizing Induced by Active Group*, https://doi.org/10.1007/978-981-96-5362-1_6

6.1 Indicator Gas Generation Pattern of Reactive Groups in Spontaneous Coal Combustion

6.1.1 Raw Coal Gas Generation Characteristics

As shown in Fig. 6.1a, in the initial stage of YM oxidative warming, the YM mass is basically unchanged, and as the temperature increases, it enters into the mass loss stage, and the rate of change of the mass at 550 °C gradually levels off. As shown in Fig. 6.1b, the WYM mass remained stable at the beginning stage of oxidative warming, and when the temperature increased to 400 °C, the sample mass began to decrease and the TG curve started the mass loss stage, which terminated at a temperature of about 650 °C; the main reason for the decrease in sample mass within the mass loss stage was the escape of gases, such as CH_4, C_2H_4, H_2O, $CO + C_2H_4$, CO_2, etc. (Lou and Jia 2019).

Figure 6.1 shows the MS curves of CH_4 (*m/z* = 16), C_2H_4 (*m/z* = 30), H_2O (*m/z* = 18), $CO + C_2H_4$ (*m/z* = 28), and CO_2 (*m/z* = 44) released from the pyrolysis process of YM and WYM. As can be seen from the figure, CH_4 gas was detected at the beginning of the coal samples and the escape was the largest at the initial temperature, and its ionic current intensity decreased with the increase of temperature, which indicated that a large amount of CH_4 gas was stored in the raw coal. After that, the intensity of CH_4 ion current decreases with increasing temperature, which is due to the fact that after the CH_4 stored in coal is released, the amount of newly generated CH_4 gas after the breakage of aliphatic hydrocarbon side chains and bridge bonds in the coal was lower than the amount of escaped gas at the stage of evaporation desorption. As shown in Fig. 6.1a, a small increase in CH_4 gas occurs at about 450 °C, which is due to the fact that the coal sample enters the combustion stage; as shown in Fig. 6.1b, a small increase in CH_4 gas occurred at ca. 480 °C, which is due to the fact that the YM ignition temperature is reached, and the fixed carbon and volatile components in the coal commenced to burn, and the temperature rose sharply, and the methyl-containing macromolecule long chains in the coal and the thick cyclic: The long chains of methyl-containing macromolecules in the coal and

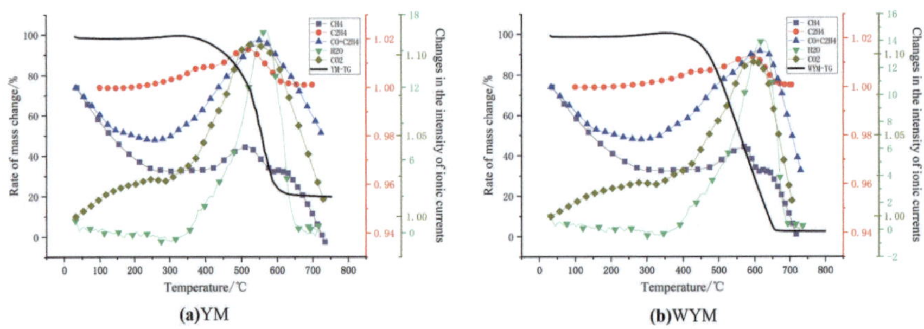

Fig. 6.1 Raw coal TG-MS plot

the thick-ringed aromatic hydrocarbons undergo a cracking reaction, releasing a large amount of CH_4 gas, so that the generation of CH_4 gas in this stage is greater than the amount of escaping gas, and the curve appears to peak. After this, the organic matter in the coal is slowly burned out, and the amount of CH_4 gas gradually decreases.

C_2H_4 corresponds to $C_2H_3^+$ which has a mass-to-charge ratio of 27, and is interfered by HCN+, so it is difficult to detect C_2H_4 and $C_2H_3^+$ accurately. As shown in the figure, the C_2H_4 gas appeared gradually at 100 °C, and the intensity of ionic flow increased slightly before 300 °C, and then continued to increase and the rate of increase increased prominently, and the intensity of ionic flow decreased and leveled off with the further increase of temperature. For WYM, as shown in Fig. 6.1b, the ion current intensity increases to the maximum value at 530 °C, which is ~1.02 times of the initial ion current intensity; the ion current intensity reaches the maximum at about 600 °C, but it is smaller than the YM ion current intensity. This is because C_2H_4 in coal is produced at a higher temperature relative to other gases, and the methyl group on the aromatic ring and with the methyl group on the aliphatic chain combine with each other after both are shed, resulting in the formation of C_2H_4.

The release pattern of H_2O is directly related to the hydroxyl group, and the curves of H_2O ion flow intensity versus temperature during the spontaneous combustion of WYM and YM are delineated in Fig. 6.1. The H_2O curves show a tendency to increase and then decrease, and the maximum ion flow intensity of H_2O occurs near the ignition temperature, and the time of occurrence is earlier than that of gases, such as CO and CO_2. After the peak, the original active groups in the coal have been consumed, and the macromolecular structure and the stable structure of the coal, such as the aromatic nucleus begin to break and cleave, competing with H_2O for energy and heat, so the ion current intensity curve of H_2O showed a decreasing trend. The maximum rate of change of the ion current intensity of the coal in YM is reached at ~550 °C, which was even lower than the temperature of WYM, and the maximum value of the ion current intensity of the coal in YM was 17 times of the initial ion current intensity, and the maximum value was 1.5 times of the initial ion current intensity. The maximum value of the ion current intensity of H_2O in YM was 17 times of the initial ion current intensity, while the maximum value of the ion current intensity of H_2O in WYM was only 14 times of the initial ion current intensity. This was because YM was less metastable than WYM and contained more reactive groups, so the oxidation reaction of YM was more swift and more intense. Since the corresponding mass-to-charge ratios of CO and C_2H_4 both 28, and both will appear in the coal pyrolysis process, the ion current intensity with a relative atomic mass of 28 was considered to be the common ion current intensity of the two gases in this MS curve, so the reason for the difference between the two curves of C_2H_4 and CO + C_2H_4 in the figure is the CO gas. As shown in Fig. 6.1a, the CO + C_2H_4 gas in YM starts to show significant release at roughly 200 °C and peaks at about 530 °C. As shown in Fig. 6.1b, the CO + C_2H_4 gas in WYM starts to show significant release at about 300 °C and peaks at 630 °C. The CO_2 gas rises with the temperature firstly with a small increase, and then rises sharply after reaching the ignition temperature of coal samples, followed by a decreasing trend. YM and WYM reached the maximum ion current intensity around 550 and 600 °C, respectively, and the maximum ion current intensity was 1.115 and 1.095 times of the initial ion current intensity, respectively. The peak of CO_2 was due to the reaction of the C–H bond of methoxy (–OCH_3) in the samples with oxygen to form the –O–C=O– bond, and due to the accumulation of the heat of the reaction and the increase of the experimental temperature, the CO_2 gas was released at about 300 °C. Owing the accumulation of the heat of the previous reaction and the increase of the experimental temperature, part of the aldehyde carbon broke the –C–O– bond connected with the ether bond –O– to produce CO and escapes, and part of the carbon connected with the ether bond breaks to produce CO_2 gas.

6.1.2 Characteristics of CH_4 Gas Generation Before and After Extraction

Figure 6.2 exhibits the MS curves of CH_4 ($m/z = 16$) released from the pyrolysis process of YM and WYM. From the figure, it can be observed that the CH_4 gas profile disclosed an overall decreasing trend with increasing temperature, but there was a relative peak of ion flow intensity during the process. The extraction treatment of the coal samples with different extractants resulted in different values of the temperature and the maximum value of the ionic flow intensity for the production of CH_4 gas. Before 380 °C, the coal samples before and after extraction had similar patterns of CH_4 gas production, and then as the temperature continued to increase, the pattern of the gas production appeared to be different, in which the original coal of YM reached the peak value of the ionic flow intensity of the gas at the highest temperature of about 530 °C, and the gas ion current intensity peak was the lowest after extracting YM with petroleum ether, and it was about 0.975 times of the initial ion current intensity. This is because petroleum ether mainly extracts the aliphatic atoms in the molecular structure of YM, so the aliphatic hydrocarbon gases produced by petroleum ether extracted coal samples were lower than those produced by raw coal and other extractants. The overall trend of CH_4 gas altered in WYM before and after the use of different extractants was similar to that of YM, but the maximum ionic flow intensity of CH_4 gas is higher than that of CH_4 gas in the coal samples of WYM after the same type of extractant was used for the extraction of YM and WYM respectively. However, after the extraction of YM and WYM with the same extractant, the maximum value of CH_4 gas ion current intensity in WYM coal samples was smaller than that in YM, and the temperature at the maximum value of ion current intensity was larger than that in YM.

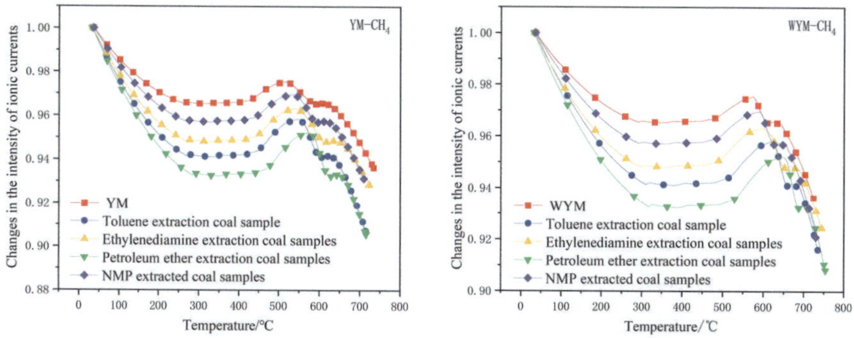

(a)YM (b)WYM

Fig. 6.2 CH_4 gas variation diagram

6.1.3 Characteristics of H_2O Gas Generation Before and After Extraction

Figure 6.3 shows the MS curves of H_2O released from the pyrolysis process of YM and WYM. As can be seen from the figure, with the increase of temperature, the H_2O change rate curve showed a parabolic-like trend, and its overall change rule remained basically the same. Owing the use of different extractants for the extraction treatment of YM and WYM, the maximum amount of H_2O produced and the corresponding temperature values were different. Before 300 °C, the changes of the ion current intensity of H_2O produced by the coal samples before and after extraction were relatively smooth, which means that the changes of the concentration of H_2O produced by the coal samples at this time were relatively smooth, and the content of H_2O produced by the samples increased expeditiously with the further elevation of the temperature. Among them, as shown in Fig. 6.3a, the YM raw coal reached the maximum H_2O production at 550 °C, which was roughly 17 times of the initial ionic current intensity. Compared with the other four coal samples, the coal sample with ethylenediamine extraction of YM was the first to reach the maximum ionic current intensity of H_2O, and it was ~7 times of the initial ionic current intensity. From Fig. 6.3b, it can be seen that the trends of H_2O ionic current intensity in WYM and YM are basically the same. However, the temperature at which the maximum ion current intensity was reached in WYM generally lagged behind that in YM, and the peak value was generally lower than that in YM, and the maximum ion current intensities of WYM raw coal and ethylenediamine-extracted coal samples were 14 and 6 times of the initial ion current intensity, respectively. This is because ethylenediamine extracts the –OH in the coal molecular structure, resulting in the reduction of –OH content, which is the main group of the formed H_2O, and thus the ionic current intensity of H_2O is the lowest in the ethylenediamine-extracted coal samples.

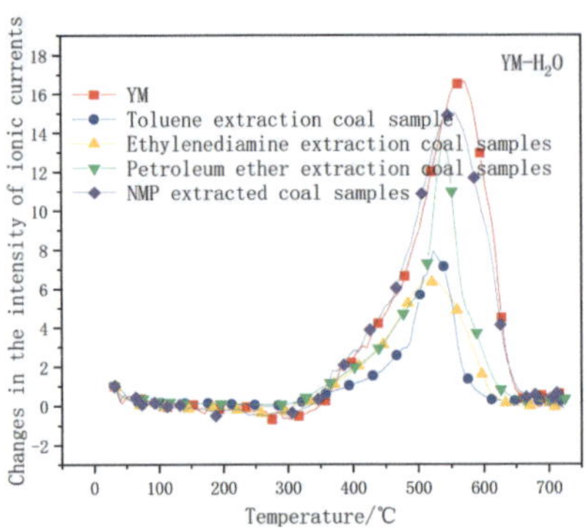

(a)YM

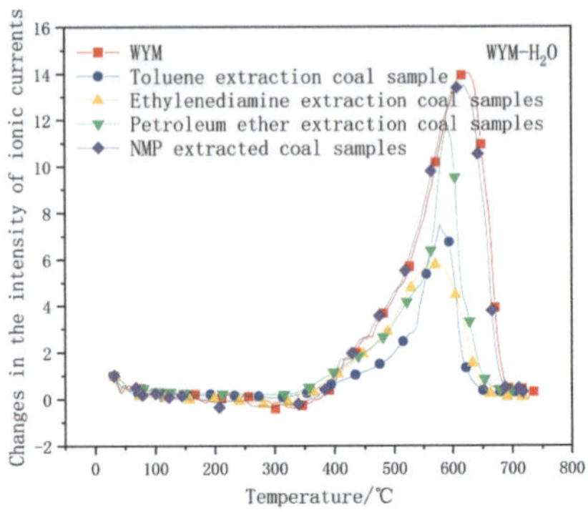

(b)WYM

Fig. 6.3 H_2O gas variation diagram

6.1.4 Study on the Thermal Effect of Coal-Oxygen Reaction in the Process of Coal Spontaneous Combustion

Figure 6.4 shows the MS curves of C_2H_4 released from the pyrolysis process of YM and WYM. As can be seen from the figure, the overall trend of the C_2H_4 gas curve in each sample is basically the same, and it activated to be generated gradually at 100 °C, and the intensity of the ion current increases quickly with the increase of temperature around 200 °C, and then finally shows a decreasing trend, falling to close to the initial value of the test. The generation of C_2H_4 gas is a complex and lengthy process, which is mainly generated by the aliphatic side chain on the aromatic ring in the structure of coal macromolecules and the free aliphatic hydrocarbons through the gradual cleavage of free radicals. The C_2H_4 gas is generated by the gradual cleavage of free radicals. The ionic strength of the C_2H_4 gas generated from the extracted coal samples was lower than that of the C_2H_4 gas generated from the original coal, among which the ionic strength of the C_2H_4 gas after petroleum ether extraction was the lowest, i.e., the concentration of C_2H_4 gas after petroleum ether extraction was the lowest. This is because petroleum ether mainly extracts aliphatic hydrocarbon functional groups in the molecular structure of coal, so the C_2H_4 gas generation in the extracted coal samples decreases, i.e., the reduction of aliphatic hydrocarbons leads to the reduction of C_2H_4 gas generation in coal. Figure 6.4a shows that the C_2H_4 gas generated from the YM raw coal reaches the maximum ion current intensity at about 520 °C, at which time the ion current intensity of C_2H_4 gas is about 1.0158 times of the initial ionic current intensity; the ionic current intensity of C_2H_4 gas from YM after petroleum ether extraction reaches the maximum value at about 550 °C, and the maximum ionic current intensity is about 1.0065 times of the initial ionic current intensity. From Fig. 6.4b, it can be seen that the maximum gas concentration of the WYM raw coal is reached only at about 590 °C, and the coal sample after petroleum ether extraction of WYM reaches the maximum gas concentration at about

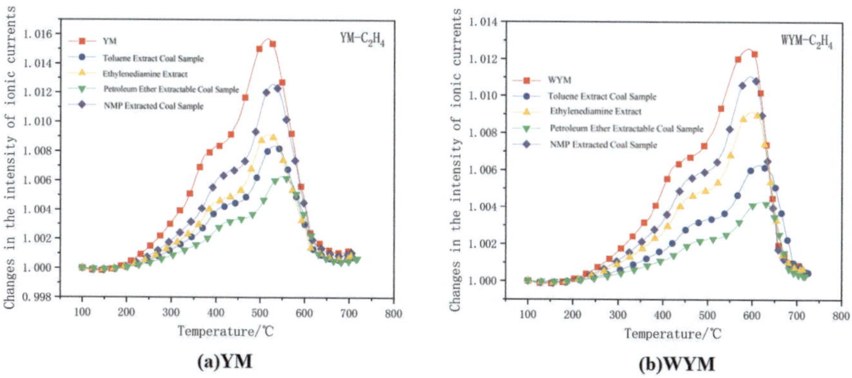

Fig. 6.4 Graph of C_2H_4 gas variations

620 °C. The temperature value at the maximum ion current intensity of the C_2H_4 gas produced by WYM lags behind that of YM, and the peak value is smaller than that of YM.

6.1.5 Characteristics of $CO + C_2H_4$ Gas Generation Before and After Extraction

Figure 6.5 shows the MS curves of $CO + C_2H_4$ ($m/z = 28$) released from the pyrolysis process of YM and WYM. From Fig. 6.5a, it can be seen that the $CO + C_2H_4$ gas curves showed a parabolic-like trend with the increase of temperature, but the use of different extractants for the extraction treatment of YM led to the different values of the temperature and the amount of $CO + C_2H_4$ gas produced. Before 300 °C, the extracted coal samples had similar patterns of $CO + C_2H_4$ gas production, and then, with the further increase of temperature, the gas production patterns appeared to be different. After that, with the further increase of temperature, the gas production pattern appeared to be different, and the YM raw coal and petroleum ether extracted YM reached the maximum value of $CO + C_2H_4$ gas concentration at 530 and 570 °C, respectively. Among them, the $CO + C_2H_4$ gas production after petroleum ether extraction of WYM was the smallest, which was only 0.975 times of the initial ionic flow intensity. From Fig. 6.5b, it can be seen that with the increase of temperature, the overall trend of the change curve of $CO + C_2H_4$ gas produced in WYM was similar to that of YM, with a relatively gentle trend in the early stage, and after 300 °C, with the further increase of temperature, the ionic current intensity of $CO + C_2H_4$ gas increased promptly, for example, the ionic current intensity of the raw coal of WYM was increased to 1.035 times of the initial ionic current intensity, which was the main reason for this was that the content of reactive groups in YM was higher than that in WYM, and the reaction was more intense. The peak of $CO +$

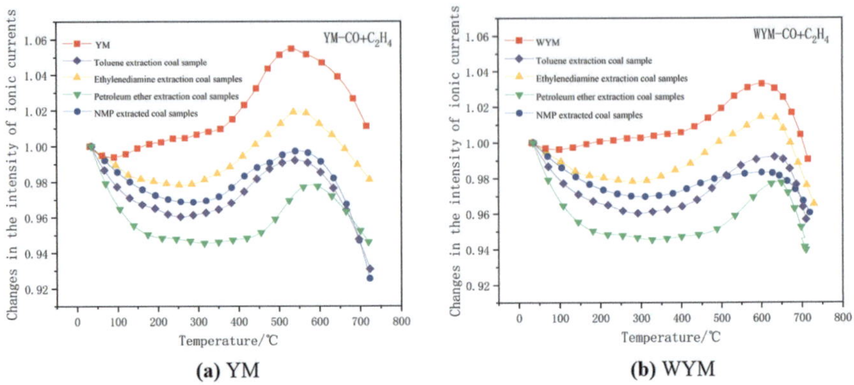

Fig. 6.5 $CO + C_2H_4$ gas change diagram

C_2H_4 during the reaction was because the oxidation reaction rate of the coal sample accelerates with the increase of temperature, and the hydroxyl groups, aliphatic hydrocarbon side chains and bridge bonds, aromatic hydrocarbons and other structures in the coal react with oxygen to produce a large number of oxygen-containing functional groups (–COOH, –C=O–, etc.), and the cracking interpretation of the oxygen-containing functional groups released CO, which made the ionic current intensity of CO increased continuously. As the reaction proceeded further and the temperature rose further, various types of irregular thick ring aromatic hydrocarbons underwent fractures and depolymerisation reactions, releasing more CO, making the intensity of the ionic current of CO increased sharply until it reached the peak.

6.1.6 Characterization of CO_2 Gas Generation Before and After Extraction

Figure 6.6 shows the MS curves of CO_2 ($m/z = 44$) released from the pyrolysis process of YM and WYM. From Fig. 6.6a, it can be seen that the CO_2 gas curves increased parabolically with the increase of temperature, but the application of different extractants to the extraction process of WYM resulted in different values of the temperature at which the CO_2 gas was produced and the amount of CO_2 gas produced. The CO_2 gas concentration of the raw coal of YM peaked at 550 °C, which was about 1.115 times of the intensity of the initial ionic flow, and the extracted coal samples The temperature required to reach the maximum CO_2 gas concentration was higher and the maximum gas concentration was larger; among them, the temperature required to reach the maximum CO_2 gas concentration of the coal sample after petroleum ether extraction of YM was the highest, and the CO_2 gas concentration was smaller than that of the original coal. This is because petroleum ether mainly extracts aliphatic hydrocarbons in the molecular structure of YM, i.e., the aliphatic hydrocarbons

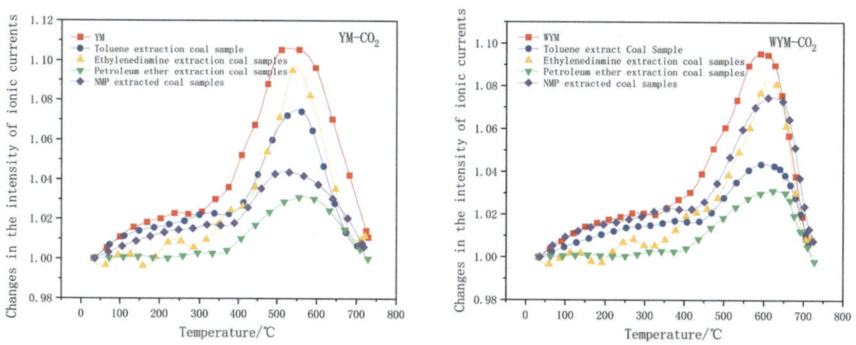

(a) YM (b) WYM

Fig. 6.6 CO_2 gas change diagram

become less after extraction, which leads to a slower oxidation reaction process of the coal. The CO_2 gas generation patterns of the original coal of WYM and the WYM extracted with the four extractants were similar to those of YM as a whole, and the only difference was the gas peak and the temperature at the peak, which were lower than the gas peak of YM and reached the peak at a higher temperature. The maximum ionic strength of CO_2 gas produced from WYM raw coal was about 1.095 times of the initial ionic flow strength. The temperature at the maximum WYM ionic flow intensity after petroleum ether extraction is about 630 °C. This is because the temperature at the maximum value of the ionic flow intensity is lower in YM than in WYM because of the lower degree of deterioration in YM and more reactive functional groups in the coal, which leads to a higher reaction rate and a faster reaction process.

6.2 Analysis of Reaction Characteristics of Coal Spontaneous Combustion Active Groups

6.2.1 Analysis of Molecular Orbital Characteristics of Active Groups in Coal Samples

In this paper, Gaussian 16 quantum chemistry software is used to optimize the model and calculate the frequency (opt + freq). Figure 6.7 shows the created model. In the calculation process, density functional theory (DFT) was selected as the method, specifically B3LYP method (Bi et al. 2024), and 6-311G(d) was selected as the basis group. The minimum point is calculated by opt + freq for the reactants and products, and the molecular orbital and orbital energy, atomic charge distribution, charge density, electrostatic potential value and thermal energy value of each compound can be found in the output file (.chk) Result (Luan et al. 2024). The highest occupied molecular orbital (HOMO) and lowest unoccupied molecular orbital (LUMO) are very important parameters in quantum chemistry.

Frontier orbital theory (Fukui 1985) holds that the occurrence of chemical reactions is due to the mutual attraction of HOMO and LUMO between the two reactants, resulting in electron transfer, which leads to the occurrence of reactions. The bonding properties of chemical reactions depend on the energy levels and charge distribution of the front orbitals. According to the theory of organic chemistry, during electrophilic reactions, electrophiles attack the place with the highest HOMO, that is, the atom with the most energetic electron. In contrast, in a nucleophilic reaction, the nucleophile attacks the atom with the lowest LUMO, the atom that can accept electrons.

Therefore, in this book, the HOMO and LUMO orbital energy levels of the constructed molecular structure of the active group were simulated, and the oxidation characteristics of the active group were investigated according to the energy gap value of the front orbital.

(a) YM Molecular Orbital Characteristics

The molecular orbital diagram of YM active groups is shown in Fig. 6.8, and the energy level distribution is shown in Table 6.1. The energy gap value (ΔE) of HOMO–LUMO reflects the transition ability of electrons from occupied orbital to empty orbital. Compared with LUMO, the larger the energy level (MOs) of HOMO, the smaller the energy gap value (ΔE) between HOMO and LUMO, indicating that the smaller the electron binding force, the easier the transition will occur, and the worse the molecular stability of the reactant. Accordingly chemical reactions are more likely to occur (Li et al. 2021).

As shown in Table 6.1, the energy gap of Ar–OH was 0.16888, the energy gap of Ar–CHO was 0.15614, the energy gap of Ar–CH_2CH_3 was 0.17673, and the energy gap of the benzene ring side chain was 0.15990. It can be seen that the energy gap value of Ar–CH_2CH_3 was the largest, and that of Ar–CHO was the smallest. According to the frontier orbital hybridisation theory, the molecular stability order of YM active groups was as follows: Ar–CHO < benzene side chain < Ar–OH < Ar–CH_2CH_3.

(b) WYM Molecular Orbital Characteristics

The molecular orbital diagram of WYM was shown in Fig. 6.9, and the energy level distribution was shown in Table 6.2. As can be seen from Table 6.2, the energy gap of Ar–OH was 0.12796, the energy gap of Ar–CHO was 0.12150, the energy gap of Ar–CH_3 was 0.13258, and the energy gap of the benzene ring side chain was 0.12777. It can be seen that the energy gap of Ar–CH_3 was the largest and that of Ar–CHO was the smallest. According to the frontline orbital hybrid theory, the molecular stability order of the WYM active groups constructed was as follows: Ar–CHO < benzene ring side chain < Ar–OH < Ar–CH_3.

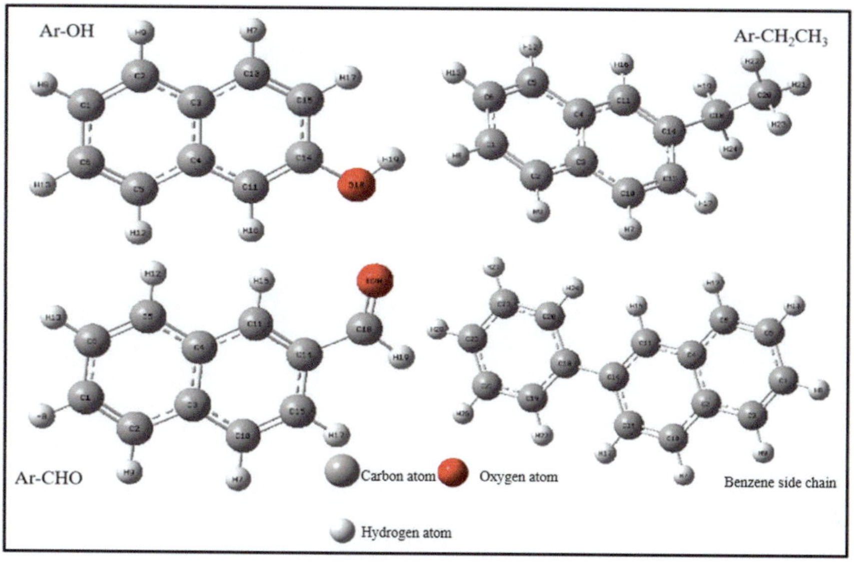

(a) Molecular structure model of YM active group

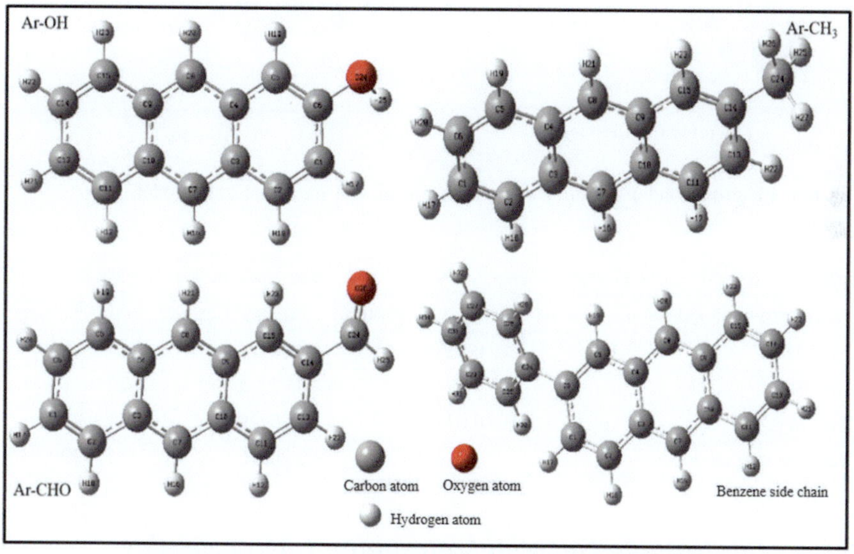

(b) WYM active group molecular structure model

Fig. 6.7 Molecular structure model of active groups in coal samples

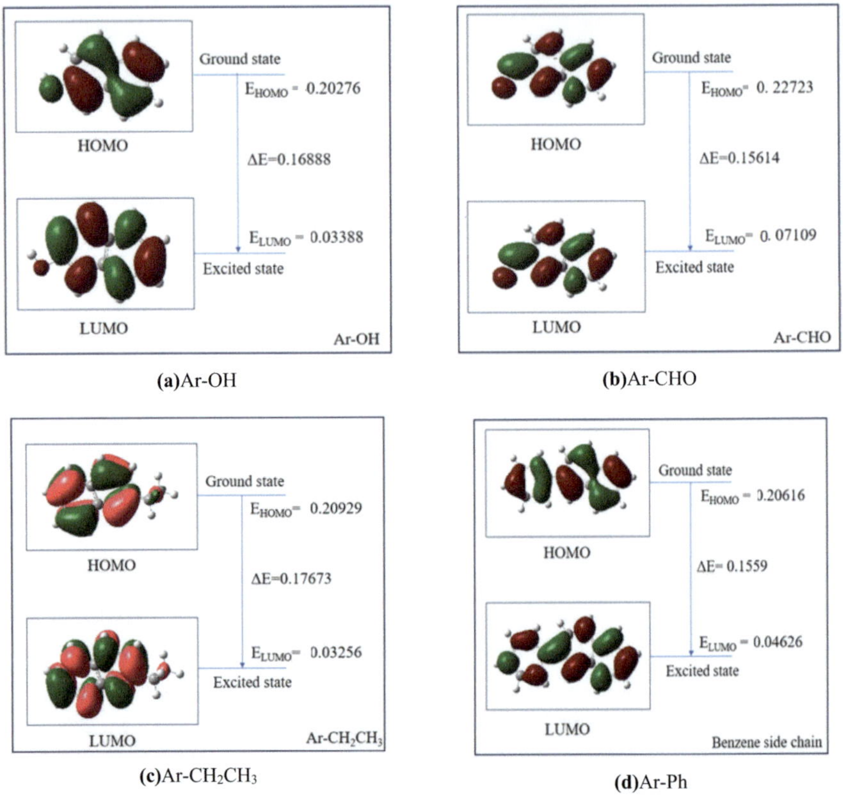

Fig. 6.8 Electron cloud distribution at the front line of YM small molecule model

Table 6.1 Frontier molecular orbital energy levels of YM active group molecular structure

Active group molecular structure name	E_{LUMO} (Hartree)	E_{HOMO} (Hartree)	ΔE (Hartree)
Ar–OH	0.03388	0.20276	0.16888
Ar–CHO	0.07109	0.22723	0.15614
Ar–CH_2CH_3	0.03256	0.20929	0.17673
Benzene side chain	0.04626	0.20616	0.15990

6.2.2 Electrostatic Potential Distribution Characteristics of Coal Sample Active Groups

In Chap. 3, the electrostatic potential of the model of coal active group molecules is analyzed, and it is found that the electrostatic potential near hydrogen atoms is positive potential, and the electrostatic potential near oxygen atoms is negative potential,

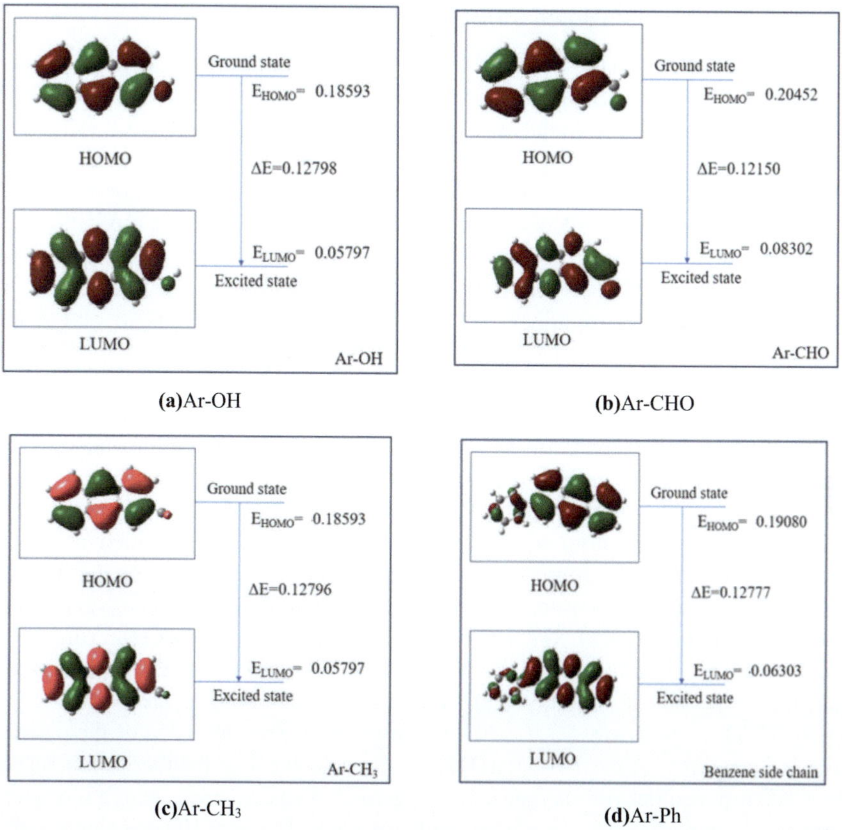

Fig. 6.9 Distribution of electron clouds in front orbit of WYM small molecule model

Table 6.2 Frontier molecular orbital energy levels of WYM active group molecular structure

Active group molecular structure name	E_{LUMO} (Hartree)	E_{HOMO} (Hartree)	ΔE (Hartree)
Ar–OH	0.05797	0.18593	0.12796
Ar–CHO	0.08302	0.20452	0.12150
Ar–CH_3	0.05701	0.18959	0.13258
Benzene side chain	0.06303	0.19080	0.12777

but this is only a direct analysis of the molecular surface. To further understand the charge size of small molecule models, Multiwfn (Zhao et al. 2024) combined with molecular visualization VMD software was used to appraise and draw the extreme value distribution of electrostatic potential on molecular surface, and the maximum and minimum values of the spatial function of electrostatic potential on molecular van der Waals surface were determined.

(a) YM Electrostatic Potential Distribution Characteristics

The projected electrostatic potential of YM active group molecular structure and the distribution of extreme points are shown in Fig. 6.10. As can be seen from Fig. 6.10a, the minimum point of –OH was on the oxygen atom (O18), with a value of −106.76 kJ/mol. This was because there was an unshared electron pair on the oxygen atom (O18) of –OH, and the electron pair produces conjugation effect with the electrons on the benzene ring and was transferred to the aromatic ring, thus presenting a negative potential. The maximum point of the potential is on the hydrogen atom (H19) connected to phenol hydroxy-oxygen (O18), with a value of 219.03 kJ/mol. In electrophilic reaction, it can be considered that the active site of the reactant was concentrated on the hydrogen atom (H19) connected to phenol hydroxy-oxygen (O18). In essence oxygen was an electrophilic reagent. The O atom was attracted to the –OH atom near the O atom and obtains an electron, the H atom in the –OH breaks away from the –OH atom, and the oxygen after obtaining an electron forms –OOH; In Fig. 6.10b, the minimum electrostatic potential point 1 was near the oxygen atom (O20) of Ar–CHO, with a value of −151.92 kJ/mol. This is because the oxygen atom was highly electronegative, while the aldehyde group is an electron-donating group, so the electron cloud is biased towards the oxygen atom under the induced action, and the negative potential is near the oxygen atom. The hydrogen atoms (H17, H7, H9, H8, H13 and H12) on the benzene ring are 6 points of the maximum electrostatic potential, which are respectively the maximum values of 1, 2, 3, 4, 5, 6 points, and the electric potential values are 84.14, 84.89, 83.89, 79.91, respectively. 77.07, 72.47 kJ/mol, the maximum value of the electric potential appears near the hydrogen (H17) on the benzene ring. In other words, on the Ar–CHO group, aldehyde-oxygen can be regarded as an active site, which is of great significance in participating in electrophilic reactions. In Fig. 6.10c, the electrostatic potential minima 1, 2 points are at the hydrogen atoms near the two ethyl groups (C18) and (C20), with values of 1.00 and 13.52 kJ/mol, that is, a hydrogen atom on $–CH_2–$ can be regarded as the active site, and electrophilic reagents can be easily pulled to the vicinity (Wang et al. 2024). In Fig. 6.10d, the points with positive electrostatic potential are all located on all hydrogen atoms connected with carbon in the benzene ring, and their values are between 54.81 and 62.76 kJ/mol; the points with negative electrostatic potential are all located inside the benzene ring, and their values are between −98.62 and −65.31 kJ/mol. According to the analysis of Fig. 3. 2d, the positive potential and negative potential in the side chain of the benzene ring tend to be in a stable state, that is, there is no active site in the side chain group of the benzene ring.

(b) WYM Electrostatic Potential Distribution Characteristics

As can be seen from Fig. 6.11a, the minimum point 1 of –OH is on the oxygen atom (O24) discerned a value of −105.35 kJ/mol, and the minimum point 2,3 is located above and below the benzene ring with values of −68.20 and −68.62 kJ/mol respectively, similar to YM. The O atom on –OH shows a negative potential. The maximum potential of 1 point is 222.46 kJ/mol on the hydrogen atom (H25) connected to phenol

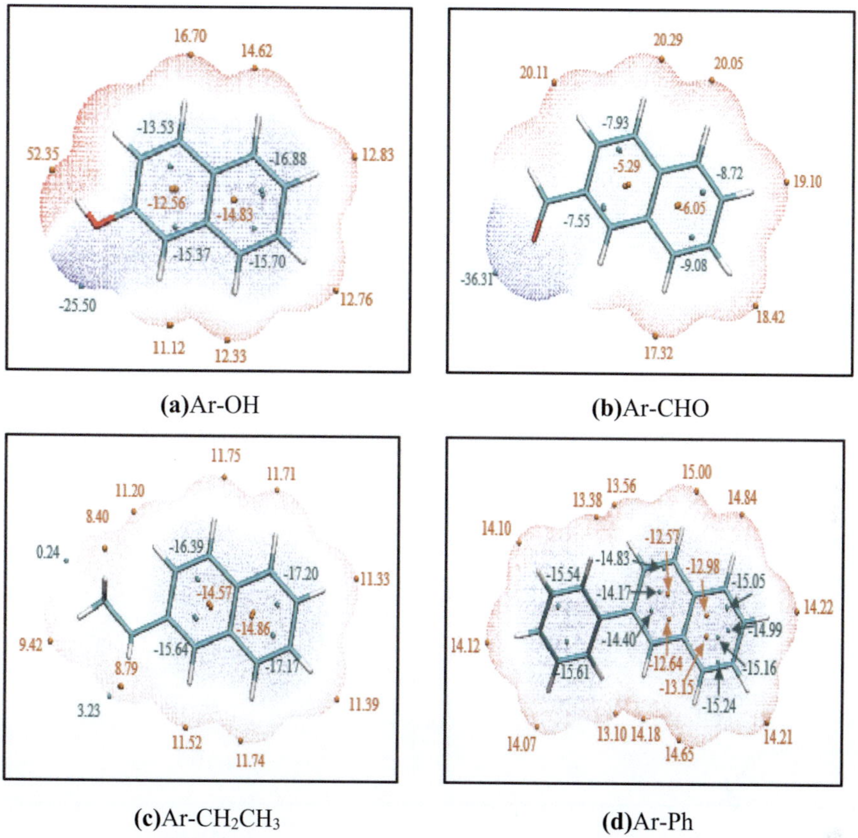

Fig. 6.10 The projected electrostatic potential of YM active group molecular structure and the distribution positions of extreme point

hydroxy-oxygen (O24). In the electrophilic reaction, it can be considered that the active site of the reactant is concentrated on the hydrogen atom (H25) connected to phenol hydroxy-oxygen (O24). In Fig. 6.11b, the minimum electrostatic potential point 1 near the oxygen atom (O26) of benzaldehyde is −153.42 kJ/mol, which is similar to YM analysis. On the Ar–CHO group, aldehyde-oxygen can be regarded as an active site, which has important significance in participating in electrophilic reactions. In Fig. 6.11c, the electrostatic potential minimum point 1 (minimum) was on the hydrogen atom of methyl group (–CH_3), with a value of 15.15 kJ/mol, that was, one hydrogen atom on methyl group is the electrophilic reaction site. It can be seen from Fig. 6.11d that the electrostatic potential of the benzene ring side chain with the molecular structure of WYM active groups was similar to that of YM. It can be concluded that there is no active site in the benzene ring side chain groups of WYM.

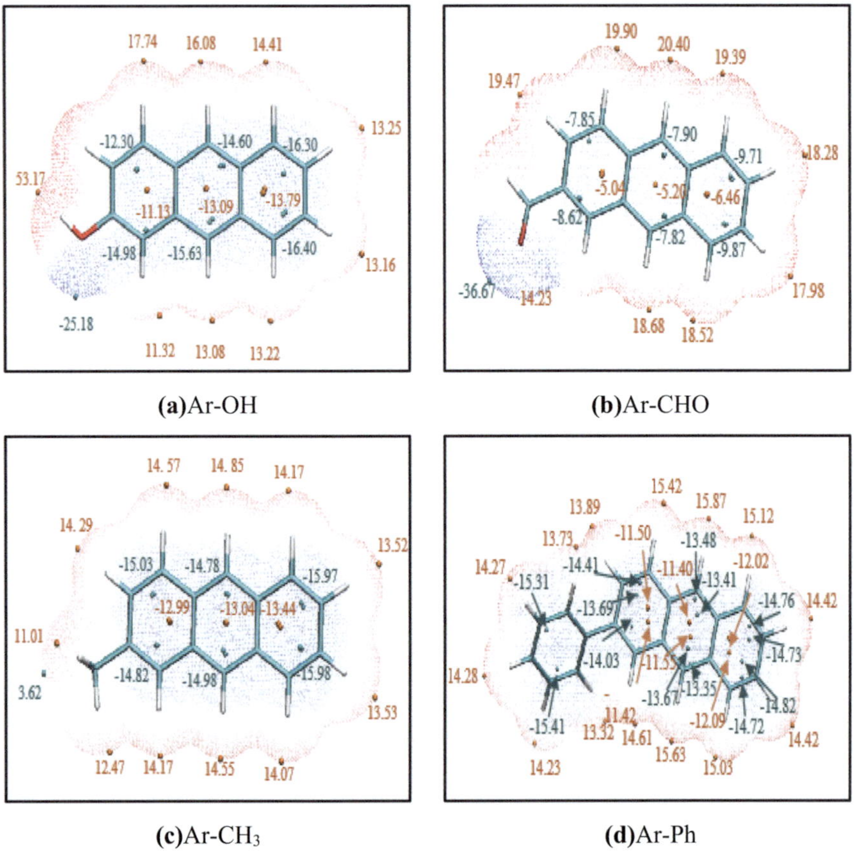

Fig. 6.11 Projection of electrostatic potential of WYM active group molecular structure model and distribution of extreme point

6.3 History of Reactive Group Oxidation

In this section, based on the judgment of the reactant active sites of YM and WYM in Sect. 6.1, the oxidation reaction history and transition state properties of the reactive groups of coal auto-ignition are analyzed by simulation calculation. According to the analysis in Sect. 6.1, it is known that the benzene ring has no active site and is not easy to be reacted, so only the oxidation reaction history and transition state properties of Ar–OH, Ar–CHO, and Ar–CH_2CH_3 groups in YM and Ar–OH, Ar–CHO and Ar–CH_3 groups in WYM were analyzed in this section. Currently, the most applied theory using quantum chemistry software is density functional theory, and the most widely used method for reaction mechanism study is the B3LYP hybridisation density functional method (Zhang et al. 2024a, b). Therefore, all the calculations in this chapter were performed using the B3LYP hybridisation flooding method with 6-311G(d) for all the basis groups; Gaussian 16 software was used to optimize the geometry of the reactants and products to obtain their stable configurations; and the TS method was used to search for the transition states between the stable configurations of the reactants and the products (Sun et al. 2023), so as to calculate each of their thermodynamic parameters. Meanwhile, to verify the reasonableness of the calculated transition states, the endowment reaction coordinates endowment reaction coordinates (IRC) were used to verify the calculated transition states (Li 2018).

Enthalpy change (ΔH) and apparent energy (ΔE) are crucial thermodynamic data that characterize the reactivity of substances (Deng et al. 2020). apparent energy (ΔE) reflects the molar energy difference between the transition state substances and the ground state energy levels of the reactant molecules in the primitive reaction, and its value is used to discriminate the difficulty of a chemical reaction; enthalpy is a pivotal characteristic parameter to characterize the reactivity of a substance, and the enthalpy change (ΔH) indicates the reaction properties of the reaction in question. According to quantum chemical calculations, enthalpy change (ΔH) is the enthalpy of the product minus the enthalpy of the reactant, and activation energy (ΔE) is the free energy of the transition state minus the free energy of the reactant, i.e., (Eqs. 6.1–6.2).

$$\Delta H = H_p - H_R \tag{6.1}$$

$$\Delta E = E_{TS} - E_R \tag{6.2}$$

where, H_P—enthalpy of products, H_R—enthalpy of reactants; E_{TS}—free energy of transition state, E_R—free energy of reactants. For any given reaction, when $\Delta H > 0$, it means that the reaction is absorptive, and when $\Delta H < 0$, it means that the reaction is exothermic. When $\Delta E < 40$ kJ/mol, the reaction is spontaneous at room temperature; when ΔE between 40 and 100 kJ/mol, the reaction can be carried out at 30–70 °C; when $\Delta E > 100$ kJ/mol, the reaction can be carried out at 70–200 °C (Li 2022).

Based on the determination of the key reactive groups in the spontaneous combustion of coal in Sect. 6.1, and the analysis and discrimination of the active sites of the reactants, and on the basis of the previous analyses and calculations of the oxidation reaction processes of the reactive groups in the oxidation process of coal by scholars, such as Wang Deming (Zhang 2012; Xu 2001; Ge and Li 2003; Wang et al. 2014), and in combination with the analyses of the reaction mechanisms in organic chemistry (Qian et al. 2015), the oxidation reaction histories of the key reactants in the YM and the WYM were hypothesized as follows:

Ym:

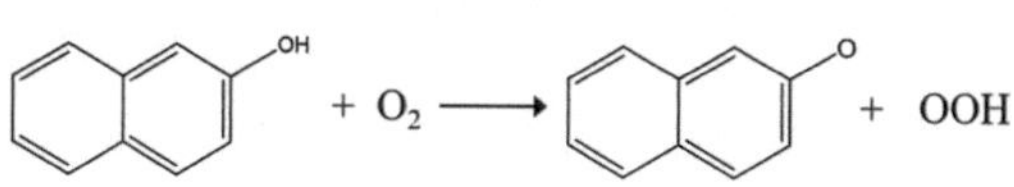

(6.3)

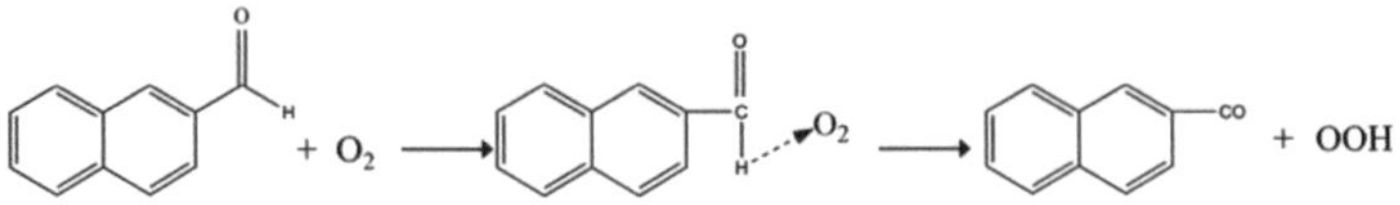

(6.4a)

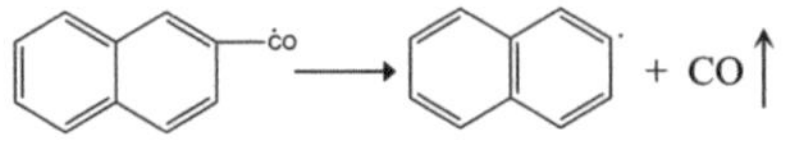

(6.4b)

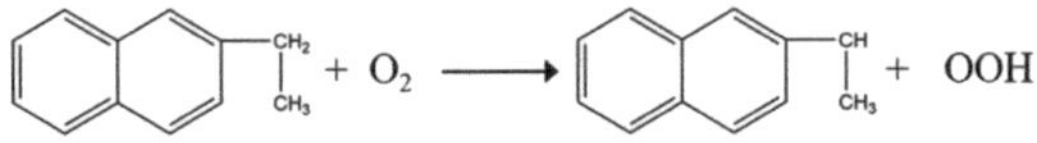

(6.5)

Wym:

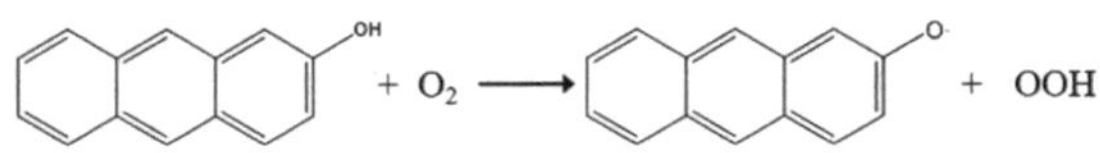

(6.6)

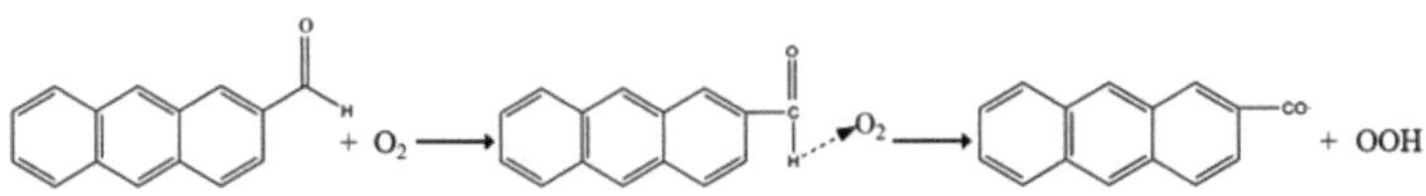

(6.7a)

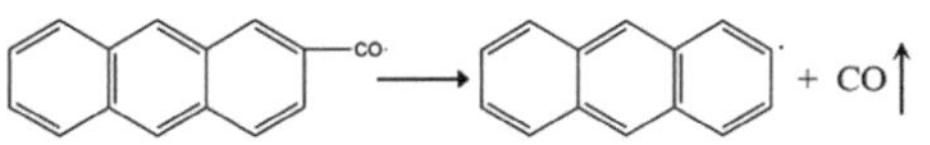

(6.7b)

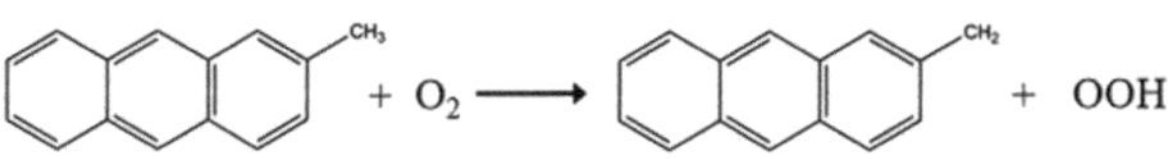

(6.8)

From reactions (Eqs. 6.3–6.8), it can be seen that the first step of the oxidation process of reactants needs to go through the process of oxygen (O_2) seizing hydrogen atoms. After Ar–CHO is seized by O_2, it will form Ar–CO-radical and –OOH radical, and after that, the C=C double bond of Ar–CO-radical will be broken, and –CO radical and benzene ring will be broken to form CO gas. After Ar–OH undergoes O_2 seizing, it will form Ar–O-radical and –OH radical; the reaction history of Ar–CH_3 is similar to that of Ar–OH, and it will form Ar–CH_2-radical and –OH radical after undergoing O_2 seizing. Ar–OH undergoes hydrogen capture by O_2 to form Ar–O– and –OOH radicals; the reaction course of Ar–CH_3 is similar to that of Ar–OH, which undergoes hydrogen capture by O_2 to form Ar–CH_2– and –OOH radicals. From the theoretical speculation of the reaction course, it was found that the first step of the reactants in the oxidation process of spontaneous combustion of coal was the process of hydrogen capture by oxygen (O_2), and after the capture of hydrogen, the –OOH radical will be formed with oxygen atoms, and due to the instability of the –O–O– bond in the –OOH radical, the –OOH radical can be directly generated by the breakage of the bond –OH radical, which is not only a reactant that triggers the cyclic oxidation, but also a generator of the –OH radical after the chemical adsorption. The –OH is not only the reactant that triggers cyclic oxidation, but also the product after chemical adsorption; the generation of –OH will cause a series of chain reactions, and then go through a complex process of changes, such as inversion. The reaction history involves multiple types of radical reactions, so this section only lists the first step of all the reactions and analyzes them with simulation calculations, and the detailed theoretical derivation of the reaction history of the reactive groups of coal auto-ignition is analyzed in Sect. 6.3. The oxidation reaction history and transition state properties of reactants in YM and WYM are analyzed computationally in the following:

6.3.1 YM Active Group Oxidation Reaction History

The reaction history of the reactants (Ar–OH, Ar–CHO, Ar–CH_2CH_3) in YM is shown in reactions (6.3)–(6.5), and the optimized molecular configurations of each major reactant R1–R3, transition states TS1–TS3, and generators P1–P3, as well as the conformational parameters (bond lengths, bond angles, and dihedral angles), as simulated by Gaussian 16 software, are listed in Table 6.3.

As shown in Table 6.3, in the course of Ar–OH oxidation reaction, under the combined effect of spatial site resistance and electron cloud in the front line orbitals, the oxygen molecules attacked the bridge bond –OH in the R2 reaction and attracted the H atoms on the –OH group, which led to the bond length of the O–H bond R (H19, O18) in the bridge bond to change from 0.97 to 1.36 Å until it was broken, and formed the –OH radicals with the oxygen atoms. The dihedral angles D (C15, C14, O18, H19) and D (C11, C14, O18, H19) change from 0° and 180° to 128.4° and 52° in the course of the R2 reaction, which indicates that the H-atoms are shifted towards O–O due to the attraction of O-atoms to H-atoms.

During the Ar–CHO oxidation reaction, when oxygen molecules attack the bridge bond –CHO in the R1a reaction, under the combined effect of spatial site resistance and the electron cloud in the front line orbitals, the oxygen molecules firstly attract the H atoms on the –CHO, which causes the bond length of the C–H bond R (C18, H19) on the bridge bond to change from 1.11 to 1.62 Å until it breaks and then forms –OH with the O atoms, thus generating Ar–CO– radicals. The C–H bond R (C18, H19) on the bridge bond is lengthened from 1.11 to 1.62 Å until it breaks and forms –OOH radical with O atoms, resulting in the Ar–CO– radical; the C–C bond R (C14, C18) on the Ar–CO– radical is lengthened from 1.36 to 2.66 Å until it breaks, resulting in the formation of the CO gas and benzene ring radical.

In the Ar–CH_2CH_3 oxidation reaction, O atoms, which are susceptible to electrophilic reactions, first attack the bridge bond –CH_2CH_3 in the R3 reactant and attract the H atoms on –CH_2–, resulting in the bond length of the C–H bond R (C18, H19) in the bridge bond being changed from 1.10 to 1.52 Å, until it breaks, and then it forms an –OH radical with the O substituent. The bond angle A (H19, C18, H24) changes from 106.45° to 94.31° in the transition state, and the bond angle A (H19, C18, C20) changes from 109.26° to 94.48° in the transition state, and due to the attraction of the O atoms, the H19 gets closer and closer to the O atoms, which leads to the decrease of the bond angle.

Figure 6.12 shows the optimized structures of the transition states of the reactions (6.3)–(6.5), where the O–O in TS1–TS3 are almost aligned along the C14–C18/O18

Table 6.3 Conformational parameters of YM

Name	Configuration	Configuration parameter		
		Bond length	Bond angle	Dihedral angles
R1		R (C15 C14) = 1.42 R (C14 C11) = 1.38 R (C14 O18) = 1.37 R (H19 O18) = 0.97	A (C14 C15 O18) = 121.38 A (C11 C14 O18) = 118.27 A (C14 O18 H19) = 108.96	D (C15 C14 O18 H19) = 0.10 D (C11 C14 O18 H19) = 179.93
TS1		R (C15 C14) = 1.48 R (C14 C11) = 1.43 R (C14 O18) = 1.28 R (H19 O18) = 1.36	A (C14 C15 O18) = 120.33 A (C11 C14 O18) = 121.62 A (C14 O18 H19) = 111.47	D (C15 C14 O18 H19) = 128.37 D (C11 C14 O18 H19) = 51.95
P1		R (C15 C14) = 1.46 R (C14 C11) = 1.44 R (C14 O18) = 1.25	A (C14 C15 O18) = 121.86 A (C11 C14 O18) = 121.30	D (C11 C14 C15 O18) = 180
R2a	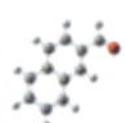	R (C15 C14) = 1.42 R (C14 C11) = 1.41 R (C14 C18) = 1.48 R (C18 H19) = 1.11 R (C18 O20) = 1.22	A (C15 C14 C18) = 119.85 A (C11 C14 C18) = 120.17 A (C14 C18 O20) = 124.90 A (C14 C18 H19) = 114.56	D (C15 C14 C18 H19) = 0 D (C15 C14 C18 O20) = 179.99 D (C11 C14 C18 H19) = 179.99 D (C11 C14 C18 O20) = 0
TS2a		R (C15 C14) = 1.42 R (C14 C11) = 1.39 R (C14 C18) = 1.47 R (C18 H19) = 1.62 R (C18 O20) = 1.19	A (C15 C14 C18) = 119.91 A (C11 C14 C18) = 119.35 A (C18 C19 O20) = 128.67 A (C14 C18 H19) = 116.30	D (C15 C14 C18 H19) = 0.02 D (C15 C14 C18 O20) = 179.99 D (C11 C14 C18 H19) = 179.98 D (C11 C14 C18 O20) = 0.01
P2a	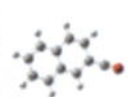	R (C15 C14) = 1.46 R (C14 C11) = 1.43 R (C14 C18) = 1.36 R (C18 O19) = 1.17	A (C15 C14 C18) = 119.28 A (C11 C14 C18) = 121.72 A (C14 C18 O19) = 179.63	D (C15 C14 C18 O19) = 0.014 D (C11 C14 C18 O19) = 179.99
TS2b	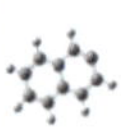	R (C15 C14) = 1.39 R (C14 C11) = 1.36 R (C14 C18) = 2.66	A (C15 C14 C18) = 116.50 A (C11 C14 C18) = 117.04 A (C14 C18 O19) = 109.12	D (C15 C14 C18 O19) = 179.95 D (C11 C14 C18 O19) = 0.12

(continued)

Table 6.3 (continued)

Name	Configuration	Configuration parameter		
		Bond length	Bond angle	Dihedral angles
P2b		R (C15 C14) = 1.40 R (C14 C11) = 1.36	A (H17 C15 C14) = 121.99 A (H16 C11 C14) = 122.80	D (H17 C15 C14 C11) = 180
R3		R (C15 C14) = 1.42 R (C11 C14) = 1.38 R (C14 C18) = 1.51 R (C18 H19) = 1.10 R (C18 H24) = 1.10 R (C18 C20) = 1.54 R (C20 H21) = 1.10 R (C20 H22) = 1.10 R (C20 H23) = 1.10	A (C11 C14 C18) = 121.59 A (C14 C18 H19) = 109.43 A (C14 C18 H24) = 109.48 A (H19 C18 H24) = 106.45 A (H19 C18 C20) = 109.26 A (H24 C18 C20) = 108.99 A (C18 C20 H22) = 110.86 A (C18 C20 H21) = 110.89 A (C18 C20 H23) = 111.21	D (C15 C14 C11 C18) = 178.69 D (C15 C14 C18 H24) = 45.80 D (C15 C14 C18 H19) = 162.11 D (C14 C18 H19 H24) = 118.21 D (C18 H19 H24 C20) = 40.67 D (C18 C20 H22 H23) = 121.81 D (C20 H21 H22 H23) = 37.64
TS3	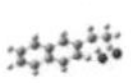	R (C15 C14) = 1.44 R (C11 C14) = 1.40 R (C14 C18) = 1.45 R (C18 H19) = 1.52 R (C18 H20) = 1.03 R (C18 C21) = 1.51 R (C21 H22) = 1.10 R (C21 H23) = 1.10 R (C21 H24) = 1.09	A (C11 C14 C18) = 119.77 A (C14 C18 H19) = 109.13 A (C14 C18 H20) = 114.23 A (H19 C18 H20) = 94.31 A (H19 C18 C21) = 94.48 A (H24 C18 C21) = 114.41 A (C18 C21 H22) = 110.99 A (C18 C21 H23) = 110.33 A (C18 C21 H24) = 112.66	D (C15 C14 C11 C18) = 178.04 D (C15 C14 C18 H20) = 167.39 D (C15 C14 C18 H19) = 91.18 D (C14 C18 H19 H20) = 116.32 D (C18 H19 C21 H22) = 23.31 D (C18 C21 H22 H23) = 120.78 D (C21 H22 H23 H24) = 37.94

(continued)

Table 6.3 (continued)

Name	Configuration	Configuration parameter		
		Bond length	Bond angle	Dihedral angles
P3		R (C15 C14) = 1.44 R (C11 C14) = 1.41 R (C14 C18) = 1.41 R (C18 H19) = 1.09 R (C18 C20) = 1.50 R (C20 H21) = 1.10 R (C20 H22) = 1.10 R (C20 H23) = 1.10	A (C15 C14 C18) = 122.85 A (C14 C18 H19) = 116.70 A (C14 C18 H20) = 126.59 A (H19 C18 C20) = 116.68 A (C18 C20 H21) = 111.87 A (C18 C20 H22) = 110.74 A (C18 C20 H23) = 112.85	D (C15 C14 C11 C18) = 179.93 D (C15 C14 C18 H19) = 177.91 D (C14 C18 H19 C20) = 177.89 D (H19 C18 C20 H21) = 85.39 D (H19 C18 C20 H22) = 33.18 D (H19 C18 C20 H23) = 153.94 D (C20 H21 H22 H23) = 38.62

Note In the diagram, the large light-colored balls are carbon atoms, the small light-colored balls are hydrogen atoms, and the red-colored balls represent oxygen atoms

direction and point to the center of the benzene ring. The distance R (O, H) between the trapped H atom and O atom is between 1.08 and 1.12 Å; the distance R (C, H) between the trapped H atom and C atom in the reactants is between 1.52 and 1.63 Å. The angle A (C, H, O) of the arrangement of the C, H, and O atoms is distributed at 172.90°–175.78°, and it can be seen that the attraction of the oxygen atom to the hydrogen atom effect makes R (C, H) longer and longer until the C–H bond is broken; while the attraction of O atoms makes the C, H, and O arrangement close to a straight line.

Table 6.4 shows the thermodynamic parameters during the reactant oxidation process in YM auto-ignition, and Fig. 6.13 shows the energy barrier diagram for each reactant oxidation process. By comparing and analyzing the reaction activation energies of R1, R2a, and R3 (Xia 2023), it can be found that the activity order of the side-chain reactive groups in the complex reaction with oxygen is –OH > –CHO > $-CH_2CH_3$.

In the analysis of the previous section, it can be seen that in the process of Ar–OH oxidation reaction, the hydroxyl bond (–OH) is mainly involved in the reaction, and the oxidation reaction firstly undergoes the process of hydrogen capture by oxygen, and the –O–H– bond on the hydroxyl group is attracted by the oxygen atom and stretched and broken, forming –OOH radicals and Ar–O-radicals, and the activation energy of the reaction of this process is 115.26 kJ/mol, and the enthalpy is 135.10 kJ/mol. The apparent activation energy of this process is 115.26 kJ/mol, and the enthalpy is 135.10 kJ/mol. ΔH of this process is >0, so it can be seen that this reaction is a heat-absorbing reaction, and it needs to absorb 135.10 kJ/mol of heat before it can take place; ΔE of this process is more than 100 kJ/mol, which means that it is difficult to take place at room temperature, and it needs a certain amount of heat

Fig. 6.12 Optimized structure of transition states for reactions R1–R3

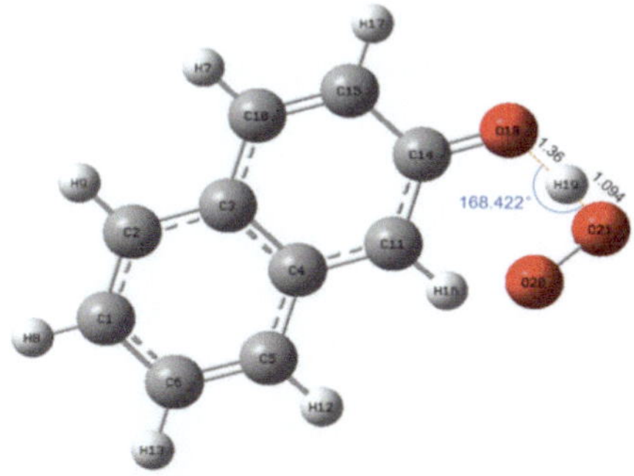

(a) Optimized structure of the transition state of TS1

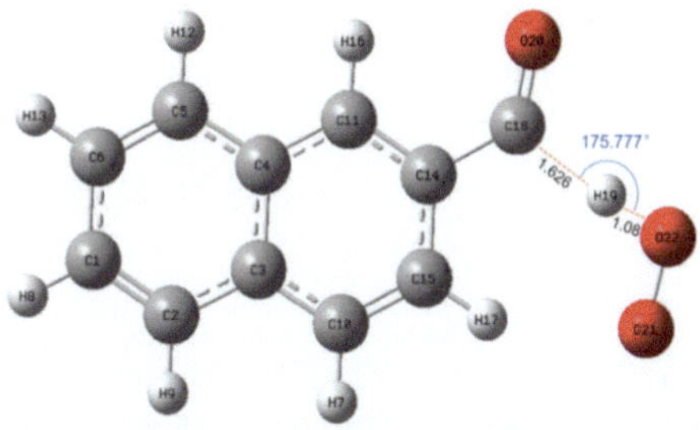

(b) Optimized structure of the transition state of TS2a

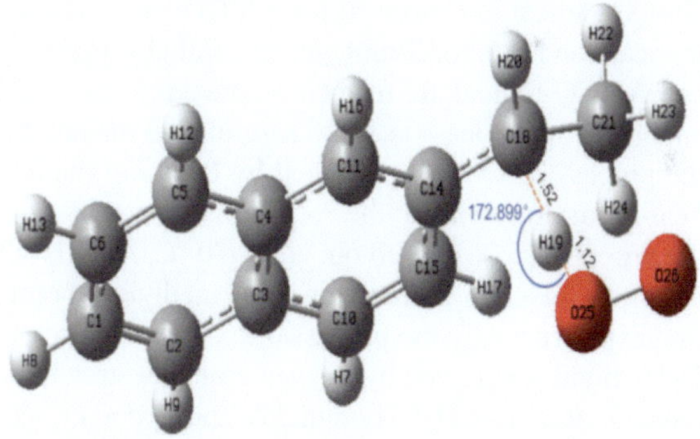

(c) Optimized structure of the transition state of TS3

to be accumulated, and it can be carried out in the temperature of 70–200 °C. In the process of Ar–CHO oxidation reaction, the aldehyde bond (–CHO) is mainly involved in the reaction, and the oxidation reaction firstly undergoes the process of hydrogen capture by oxygen, the –C–H– bond on the aldehyde group is attracted by the oxygen atom and stretches and breaks, and forms –OOH and Ar–CO-radicals, the activation energy of the reaction in this process is 177.22 kJ/mol, and the enthalpy

Table 6.4 Thermodynamic parameters of R1–R3 reaction process

Typology	Gibbs free energy, ΔE (Hartree)	Enthalpy, ΔH (Hartree)	Activation energy, ΔE (kJ/mol)	Enthalpy change, ΔH (kJ/mol)
R1	611.33	611.26	115.26	135.10
TS1	611.28	611.23		
P1	611.28	611.21		
R2a	649.43	649.36	177.22	208.32
TS2a	649.36	649.31		
P2a	649.36	649.28		
R2b	498.45	498.40	86.96	59.17
TS2b	498.42	498.37		
P2b	498.43	498.36		
R3	614.69	614.62	181.69	160.56
TS3	614.623	614.57		
P3	614.63	614.56		

Note 1 Hartree = 2625.5 kJ/mol

change is 208.32 kJ/mol. The ΔH of this process is greater than 0, so the process of the aldehyde group being captured by oxygen is a heat-absorbing process; the ΔE of this process >100 kJ/mol, which means that this process is a non-spontaneous reaction at room temperature, and it can be carried out at 70–200 °C. The C–C bond on the Ar–CO-radical is broken to form a CO gas, and the apparent activation energy of the reaction is 86.96 kJ/mol, and the enthalpy is 59.17 kJ/mol, so the Ar–CO-radical and Ar–CO-radical are formed by breaking the C–C bond. The apparent activation energy of this process is 86.96 kJ/mol, and the enthalpy is 59.17 kJ/mol. Therefore, the process of CO gas produced by breaking the bond of Ar–CO radical is also a heat-absorbing process; the ΔE of this process is 40–100 kJ/mol, which means that the process can be carried out at 30–70 °C. Ar–CH_2CH_3 oxidation reaction process, mainly involved in the reaction is the methylene bond (–CH_2–), oxidation reaction is first experienced by the process of oxygen capture of hydrogen, methylene on the –C–H– bond is attracted by oxygen atoms to stretch fracture, the formation of –OOH radicals and Ar–CH_2CH_3-radicals, the apparent activation energy of the reaction of the process for the 181.69 kJ/mol, enthalpy change to The apparent activation energy of this process is 181.69 kJ/mol, and the enthalpy change is 160.56 kJ/mol. This reaction is a heat-absorbing reaction, and it needs to absorb 160.56 kJ/mol of heat before the reaction can take place; the ΔE of this process is >100 kJ/mol, which means that the reaction can also be carried out under 70–200 °C.

From Fig. 6.13a, it can be seen that the dehydrogenation of hydroxyl group during Ar–OH reaction to generate Ar–O– and –OOH radicals needs to break through the energy barrier of 115.26 kJ/mol, which is a non-spontaneous reaction (Zhang et al. 2020). The dehydrogenation of aldehyde group in Ar–CHO needs to break through the energy barrier of 177.22 kJ/mol to form Ar–CO– and –OOH radicals, which is a

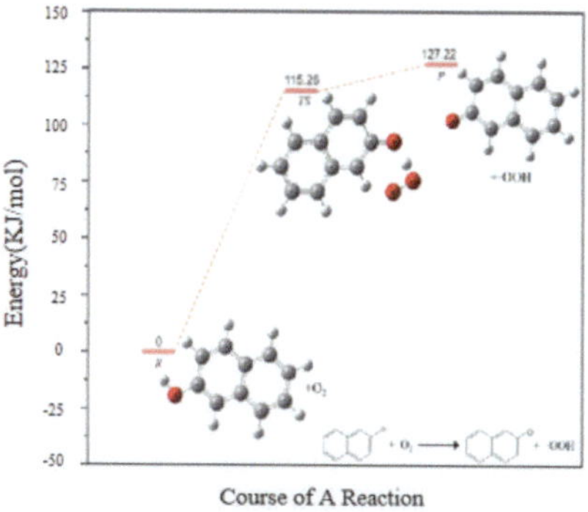

(a) R1 Reaction Energy Barrier Diagram

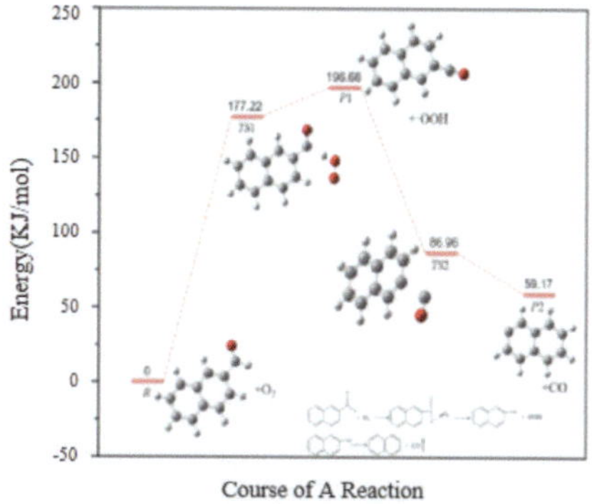

(b) R2 Reaction Energy Barrier Diagram

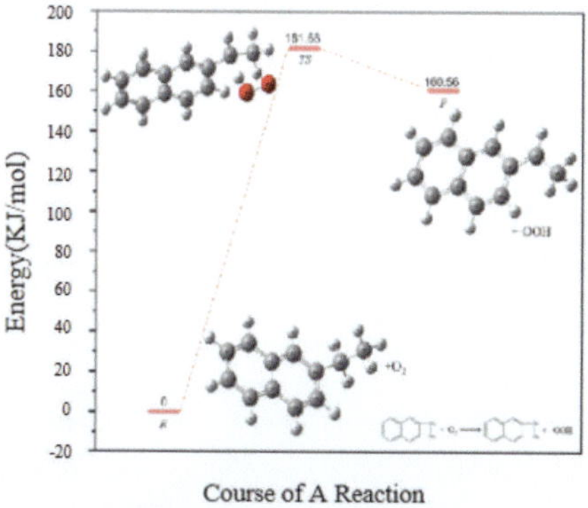

(c) R3 reaction energy barrier diagram

Fig. 6.13 YM reactive group reaction energy barrier diagram

non-spontaneous reaction; and the dehydrogenation of aldehyde group in Ar–CHO needs to break through the energy barrier of 177.22 kJ/mol to form Ar–O– and –OOH radicals, which is a non-spontaneous reaction. And –OOH radicals, and this process is non-spontaneous; the chain breaking of Ar–CO-radical to generate CO gas requires breaking the energy barrier of 86.96 kJ/mol, and this process is also a non-spontaneous reaction. The reaction course of methylene dehydrogenation in Ar–CH_2CH_3 requires breaking the energy barrier of 181.69 kJ/mol, and the formation of Ar–CH_2CH_3 radicals and –OOH radicals, and this process is a non-spontaneous reaction. TS1–TS3 The IRC reaction paths of the transition states were calculated at the 6-311G(d) basis group level by selecting IRC as the work type in the Gaussian View submission calculation interface; other parameters: IRC = (RCFC, Max Points = 10, Recorrect = Never) indicates that the transition states were centered (Zhao et al. 2024), and 10 steps were taken before and after, and the IRC reaction path energy diagrams were obtained by calculation. The IRC reaction path energy diagram is shown in Fig. 6.14.

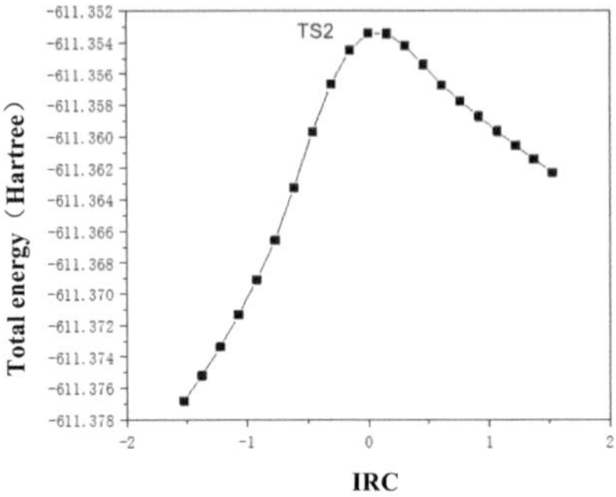

(a) R1 Reaction Process

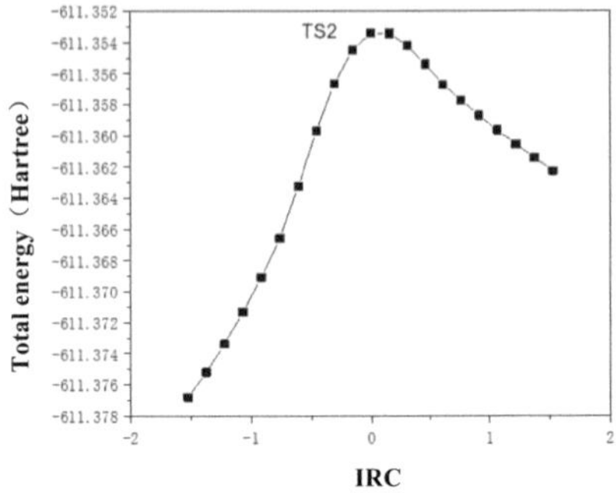

(b) R2 Reaction Process

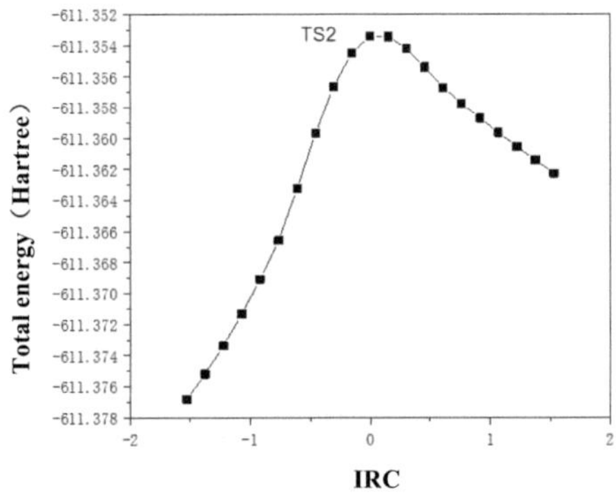

(c) R3 Reaction Process

Fig. 6.14 Variation of reaction potential energy along the reaction coordinate (IRC) for R1–R3 oxidation reaction process

As can be seen from Fig. 6.14, the energies of the R1–R3 transition state TSs were all maxima, which showed that the transition state structure was exceptional unstable, and it would be involved in the reaction immediately after its formation, and the frequency optimization of the TSs had only one imaginary frequency, and the substances that go to the two ends from the various saddle points are also all in agreement with the reactants and products written above (Lu et al. 2017), which proves the reasonableness of the transition states found while also verifying that the previously The inferred active sites, bond breaking, and reaction directions are consistent with the theory.

6.3.2 *Reactive Group Oxidation Reaction History of WYM*

The reaction history of the reactants (–OH, –CHO, and $-CH_3$) in WYM is shown in reactions (6.6)–(6.8), and the optimized molecular configurations as well as the conformational parameters (bond lengths, bond angles, and dihedral angles) of each of the major reactants R4–R6, transition states TS4–R6, and products P4–P6, which were calculated from simulations with the Gaussian 16 software, are shown in Table 6.5.

As shown in Table 6.5, in the course of Ar–OH oxidation reaction, under the combined effect of spatial site resistance and electron cloud in the front line orbitals, the oxygen molecules attacked the bridge bond –OH and attracted the H atoms on the –OH group in the R5 reaction, which made the bond length of the O–H bond R (O24, H25) in the bridge bond increase from 0.97 to 1.31 Å until it was broken, and formed –OOH with the oxygen atoms. Radical with the oxygen atom. During Ar–CHO oxidation, when oxygen molecules attack the bridge bond –CHO in the R4a reaction, under the combined effect of spatial site resistance and the electron cloud in the front line orbitals, the oxygen molecules firstly attract the H atoms on the –CHO, which causes the bond length of the C–H bond R (C24, H25) on the bridge bond to change from 1.11 to 1.63 Å until it breaks, and then it forms the –OH radical with the O atoms, and produces the Ar–CHO radical. radicals with O atoms and produces Ar–CO– radicals; The C–C bond R (C14, C24) on the Ar–CO-radical lengthens from 1.35 to 2.66 Å until it breaks, resulting in the formation of a CO gas and a benzene ring radical. In the Ar–CH3 oxidation reaction, O atoms, which are prone to electrophilic reactions, first attack the bridge bond –CH3 in the R6 reactant and attract the H atoms on –CH3, causing the C–H bond R (C24, H27) in the bridge to grow from 1.10 to 1.56 Å in length until it breaks, and then forms the –OOH radical with the oxygen atom. The bond angle A (C14, C24, H27) changes from 111.29° to 99.22° in the transition state, and it can be seen that due to the attraction of the O atoms, H27 gets closer and closer to the O atoms, which makes the bond angle smaller. R6 During the reaction, the dihedral angles D(C13, C14, C24, H25) and D(C13, C14, C24, H26) change from 59.51° to 88.15° and 88.15° to 180°, respectively. The dihedral angles D(C13, C14, C24, H25) and D(C13, C14, C24, H26) change from 59.51° and 180° to 88.15° and 88.15° respectively during the R6 reaction. It can be seen that due to the breakage of the H27 atoms, the H25 and H26 atoms are evenly distributed at the ends of the C atoms which are attached to the benzene ring due to the interacting forces and are more tightly connected to the C atoms.

Figure 6.15 demonstrates the optimized structures of the transition states of the reactions (6.6)–(6.8), where O–O in TS4–TS6 are aligned along the C14–C24/O24 direction and point to the center of the benzene ring. The distance R(O, H) between the trapped H atom and O atom is between 1.08 and 1.12 Å; the distance R(C, H) between the trapped H atom and the C atom in the reactants is between 1.31 and 1.63 Å, and the angle A(C, H, O) of the arrangement of the C, H, and O atoms is distributed at 168.97–175.64°,which is seen to show that the attraction of O atoms

Table 6.5 Conformational parameters of WYM

Name	Configuration	Configuration parameter		
		The bond length/Å	The bond angle/°	Dihedral angles/°
R5		R (C1 C6) = 1.43 R (C6 O24) = 1.37 R (O24 H25) = 0.97	A (C1 C6 O24) = 120.83 A (C6 O24 H25) = 109.05	D (C1 C6 O24 H25) = 0.01
TS5		R (C13 C14) = 1.45 R (C14 O24) = 1.29 R (O24 H25) = 1.31	A (C13 C14 O24) = 119.87 A (C14 O24 H25) = 111.90	D (C13 C14 C15 O24) = 179.61 D (C13 C14 O24 H25) = 131.05
P5		R (C1 C6) = 1.47 R (C6 O24) = 1.25	A (C1 C6 O24) = 121.94	D (C1 C6 C5 O24) = 180
R4a		R (C13 C14) = 1.43 R (C14 C24) = 1.48 R (C24 H25) = 1.11 R (C24 O26) = 1.22	A (C13 C14 C24) = 119.89 A (C14 C24 H25) = 114.50 A (C14 C24 O26) = 125.00 A (H25 C24 O26) = 120.50	D (C13 C14 C15 C24) = 179.99 D (C13 C14 C24 H25) = 0.076 D (C13 C14 C24 O26) = 179.92 D (C14 C24 H25 O26) = 179.85
TS4a		R (C13 C14) = 1.43 R (C14 C24) = 1.47 R (C24 H25) = 1.63 R (C24 O26) = 1.20	A (C13 C14 C24) = 119.99 A (C14 C24 H25) = 116.34 A (C14 C24 O26) = 128.76 A (H25 C24 O26) = 114.90	D (C13 C14 C24 H25) = 179.99 D (C13 C14 C24 O26) = 179.99 D (C14 C24 H25 O26) = 180.00
P4a		R (C13 C14) = 1.46 R (C14 C24) = 1.35 R (C24 O25) = 1.17	A (C13 C14 C24) = 119.0 A (C14 C24 O25) = 179.56	D (C13 C14 C15 C24) = 180 D (C13 C14 C24 O25) = 0.01
TS4b		R (C13 C14) = 1.41 R (C14 C24) = 2.66	A (C13 C14 C24) = 116.42 A (C14 C24 O25) = 109.17	D (C13 C14 C24 O25) = 179.93
P4b	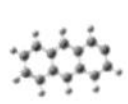	R (C13 C14) = 1.41 R (C14 C15) = 1.35 R (C13 H22) = 1.09	A (C13 C14 C15) = 126.32 A (C14 C13 H22) = 121.72	D (C13 C14 C15 H22) = 180

(continued)

Table 6.5 (continued)

Name	Configuration	Configuration parameter		
		The bond length/Å	The bond angle/°	Dihedral angles/°
R6		R (C13 C14) = 1.43 R (C14 C24) = 1.51 R (C24 H25) = 1.10 R (C24 H26) = 1.09 R (C24 H27) = 1.10	A (C13 C14 C24) = 119.41 A (C14 C24 H26) = 111.51 A (C14 C24 H25) = 111.28 A (C14 C24 H27) = 111.29	D (C13 C14 C24 H25) = 59.511 D (C13 C14 C24 H26) = 179.99 D (C13 C14 C24 H27) = 59.50 D (C24 H25 H26 H27) = 38.24
TS6		R (C13 C14) = 1.45 R (C14 C24) = 1.42 R (C24 H25) = 1.09 R (C24 H26) = 1.09 R (C24 H27) = 1.58	A (C13 C14 C24) = 120.74 A (C14 C24 H25) = 119.11 A (C14 C24 H26) = 118.93 A (C14 C24 H27) = 99.22	D (C13 C14 C24 H25) = 14.75 D (C13 C14 C24 H26) = 165.65 D (C13 C14 C24 H27) = 92.10 D (C24 H25 H26 H27) = 56.96
P6		R (C13 C14) = 1.44 R (C14 C24) = 1.48 R (C24 H25) = 1.09 R (C24 H26) = 1.09	A (C13 C14 C24) = 119.93 A (C14 C24 H26) = 121.25 A (C14 C24 H25) = 121.25	D (C13 C14 C24 H25) = 88.15 D (C13 C14 C24 H26) = 88.15 D (C14 C24 H25 H26) = 176.43

to the H atom effect makes R (C, H) longer and longer until the C–H bond is broken; while the attraction of O atoms makes C, H and O arrangement close to a straight line.

Table 6.6 shows the thermodynamic parameters during the oxidation of reactants in WYM auto-ignition, and Fig. 6.16 shows the energy barrier diagrams of the oxidation process of each reactant. By comparing and analyzing the reaction activation energies of R4, R5a, and R6, it can be found that the activity order of WYM side-chain reactive groups in the complex reaction with oxygen is –OH > –CHO > $–CH_3$.

According to the analyzis of YM in the previous section, in the process of Ar–OH oxidation reaction, the hydroxyl bond (–OH) was mainly involved in the reaction, and the oxidation course of WYM was similar to that of YM, in which the oxidation reaction firstly generates the –OOH radicals and Ar–O-radicals, and the activation energy of the reaction of the process was 100.81 kJ/mol, and the enthalpy change was 113.11 kJ/mol. The activation energy of this process was 100.81 kJ/mol and the enthalpy was 113.11 kJ/mol, which was greater than 0. It can be seen that this reaction was a heat-absorbing reaction, and it needs to absorb 113.11 kJ/mol to take place; the ΔE of this process was more than 100 kJ/mol, which means that this process can be carried out under 70–200 °C.

Fig. 6.15 Optimized structure of transition states for reactions R4–R6

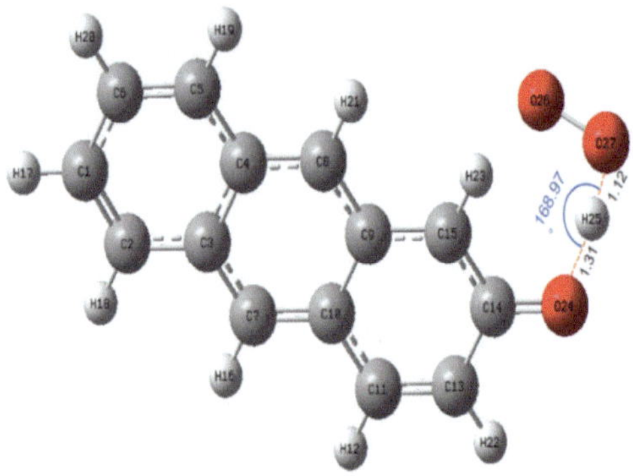

(a) Optimized Structure of The Transition State of TS4

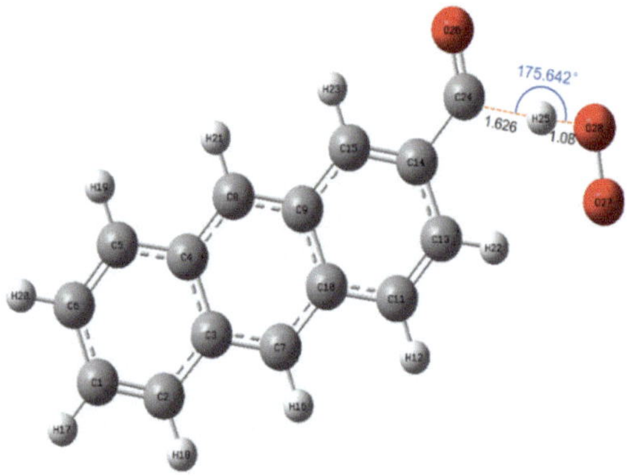

(b) Optimized structure of the transition state of TS5a

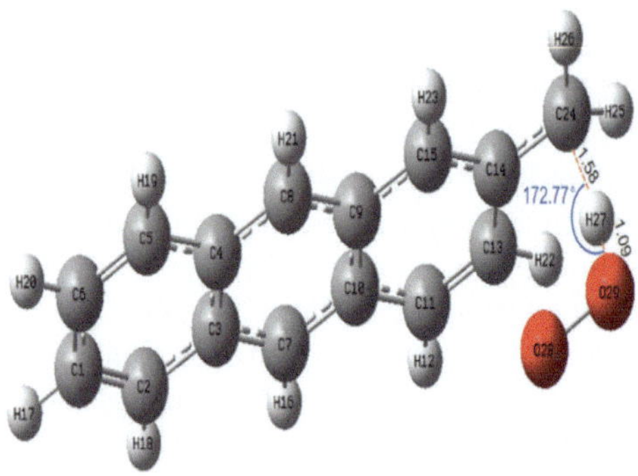

(c) Optimized structure of the transition state of TS6

Table 6.6 Thermodynamic parameters of R4–R6 reaction process

Typology	Gibbs free energy, ΔE (Hartree)	Enthalpy, ΔH (Hartree)	Activation energy, ΔE (kJ/mol)	Enthalpy change, ΔH (kJ/mol)
R4	−764.92	−764.85	100.81	113.11
TS4	−764.88	−764.83		
P4	−764.88	−764.81		
R5a	−802.96	−802.95	163.19	209.12
TS5a	−802.90	−802.90		
P5a	−802.94	−802.88		
R5b	−652.03	−651.99	48.40	114.48
TS5b	−652.01	−651.95		
P5b	−652.02	−651.95		
R6	−729.00	−728.93	180.33	232.31
TS6	−728.93	−728.87		
P6	−728.91	−728.84		

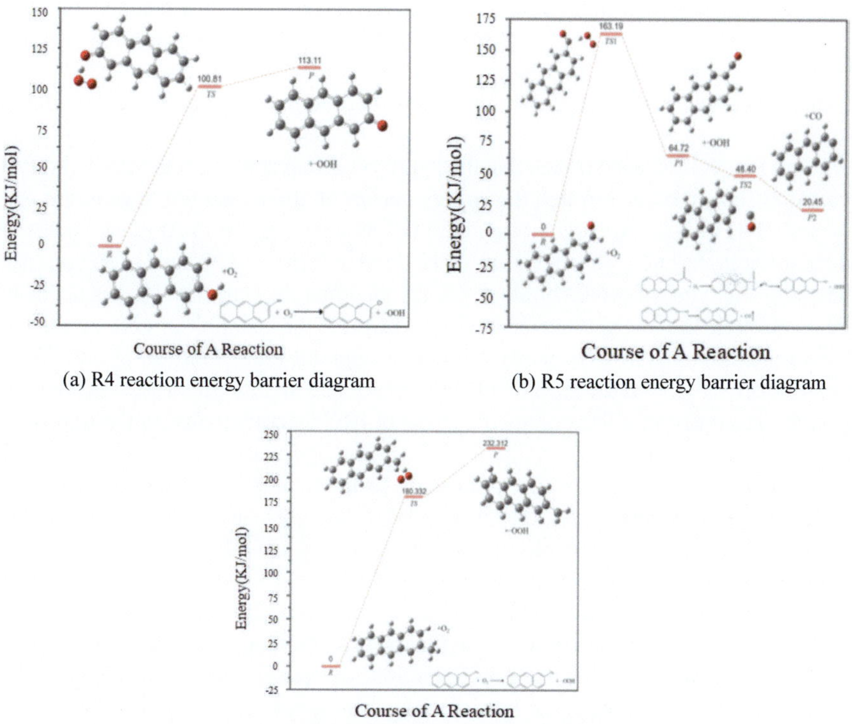

(a) R4 reaction energy barrier diagram

(b) R5 reaction energy barrier diagram

(c) R6 reaction energy barrier diagram

Fig. 6.16 Reaction energy barrier diagram for WYM reactive groups

In the oxidation of Ar–CHO, similar to the YM oxidation reaction, the main participant is the aldehyde bond (–CHO), and the oxidation reaction firstly generates –OOH and Ar–CO-radicals, the activation energy of this process is 163.19 kJ/mol, and the enthalpy change is 209.12 kJ/mol. The ΔH of the process is greater than 0, so the process of aldehyde radicals being captured by oxygen needs to absorb some heat to take place; the $\Delta E > 100$ kJ/mol of this process indicates that the process can be carried out at 70–200 °C. The activation energy of this process is 48.40 kJ/mol, and the enthalpy is 114.48 kJ/mol. The ΔH of this process is greater than 0, so the Ar–CO radical breaks the C–C bond to form CO gas, and the activation energy is 48.40 kJ/mol, and the enthalpy is 114.48 kJ/mol. –CO– radical breaks the bond to produce CO gas needs to absorb a certain amount of heat to be carried out, the ΔE of this process is between 40 and 100 kJ/mol, indicating that the process can be carried out at 30–70 °C.

In the process of Ar–CH_3 oxidation reaction, similar to the YM oxidation reaction, the main participant in the reaction is the methyl bond (–CH_3), the oxidation reaction will first generate –OOH radicals and Ar–CH_2-radicals, the activation energy of this process is 180.33 kJ/mol, the enthalpy change is 232.31 kJ/mol, the ΔH of the process is greater than 0, which shows that this reaction is a heat-absorbing reaction, and this reaction needs to absorb 232.31 kJ/mol to occur the reaction. The reaction requires 232.31 kJ/mol of heat to be absorbed before the reaction can take place; the ΔE of this process is >100 kJ/mol, which means that the reaction can be carried out at 70–200 °C.

From Fig. 6.16, it can be seen that the hydroxyl group dehydrogenation in Ar–OH reaction needs to break through the energy barrier of 100.81 kJ/mol to generate Ar–O– and –OH radicals; the dehydrogenation of aldehyde group in Ar–CHO needs to break through the energy barrier of 163.19 kJ/mol to form Ar–CO– and –OH radicals; the chain-breaking of Ar–CO radical breaks the chain to generate CO gas needs to break through the energy barrier of 48.40 kJ/mol; the reaction course of methyl dehydrogenation in Ar–CH_3 needs to break through the energy barrier of 180.33 kJ/mol to form Ar–CH_2-radicals and –OOH radicals, and all of these processes are non-spontaneous reactions. The energy diagrams of IRC reaction paths for the transition states of TS4–TS6 are shown in Fig. 6.17. From the figure, it can be seen that the IRC calculations of WYM are similar to those of YM, which proved the reasonableness of the transition states found, and also verified the consistency of the previously inferred active sites, bond breaks, and reaction directions with the theory.

Based on the theoretical analysis of the presumption of spontaneous combustion of coal micro-reaction process, through the simulation of this section of the analysis of the verification of the simulation calculations can be seen, in the process of spontaneous combustion of coal oxidation reaction, oxygen in the role of electrostatic potential, would be active to the benzene ring around the reactive group, the first to capture the active group on the electron cloud was easy to shed hydrogen atoms, the formation of –OH radicals; and aldehyde in the occurrence of the process of capturing the hydrogen, the formation of CO gas. These reactions were non-spontaneous and require a certain amount of energy accumulation to occur.

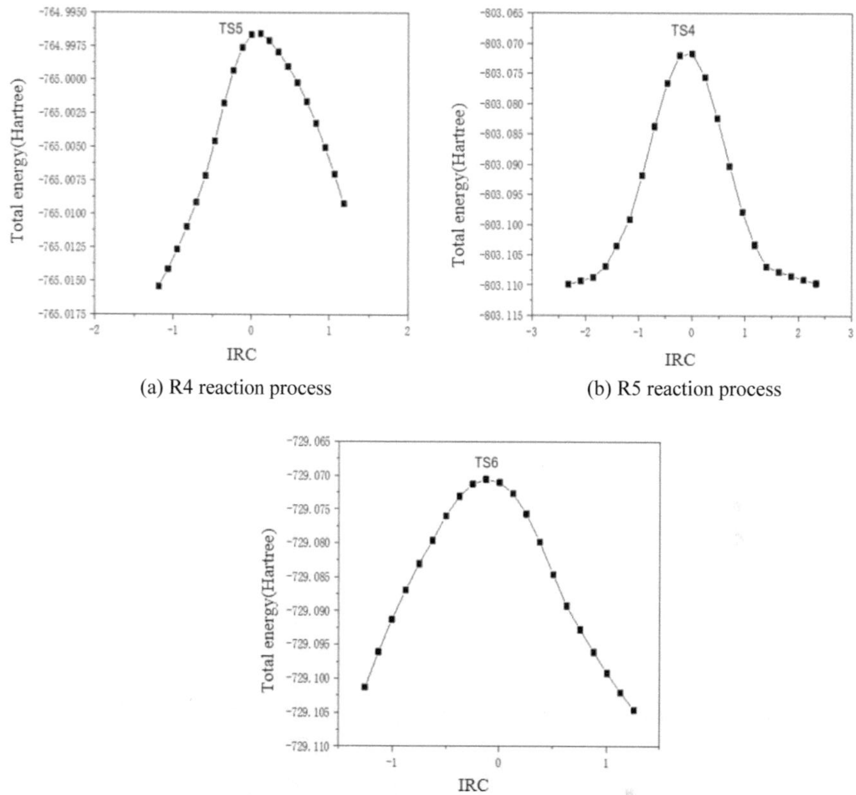

(a) R4 reaction process

(b) R5 reaction process

(c) R6 reaction process

Fig. 6.17 Variation of reaction site energy along reaction coordinates (IRC) for R4–R6 oxidation reaction process

6.4 Mechanisms Linking Coal Auto-Ignition Reactive Groups to Indicator Gases

As analyzed above, both YM and WYM, the reactive groups within their molecules will be taken away by oxygen when reacting with oxygen to form the highly oxidative –OOH by taking away the hydrogen atoms on the reactive group, which itself became a reactive radical lacking a hydrogen atom (Li et al. 2024), while –OOH was more likely to split further to become the more reactive –OH. This implies that the chain reaction of the late stage of coal auto-ignition occurred between the reactive radicals that lose the hydrogen atoms between the reactive radicals and –OH (Zhao 2023). Based on this, the Gaussian 16 simulation and calculation software was used to build and optimize the structure of the reactive radicals in YM and WYM after the loss of hydrogen atoms and further calculate their frequencies and energies (Guo et al. 2023), and the results of the calculations are shown in Table 6.7.

Table 6.7 Optimization results of coal major group analysis

Serial number	Molecular structure	H/(Hartree)	G/(Hartree)	Serial number	Molecular structure	H/(Hartree)	G/(Hartree)
1		385.737166	385.776109	16		652.640298	652.690344
2	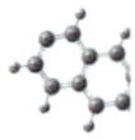	385.062707	385.102243	17	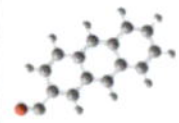	651.999771	652.050546
3	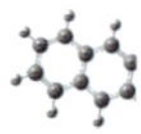	424.389694	424.432119	18	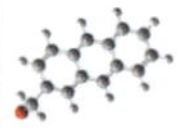	653.159203	653.211703
4	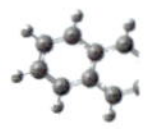	500.223839	500.269435	19	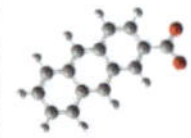	727.225160	727.278596
5		499.051357	499.095241	20	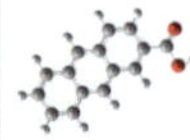	727.881624	727.934016
6		498.410868	498.455478	21		538.861750	538.910714
7		499.574414	499.620473	22	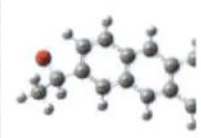	692.450357	692.505528
8	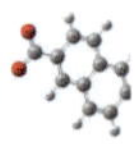	573.636441	573.683570	23		39.804411	39.828289
9	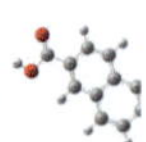	574.292965	574.339161	24		113.301116	113.323559
10	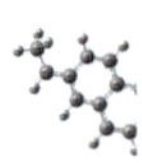	463.677562	463.725293	25		188.565756	188.590067
11	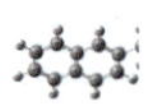	539.510927	539.559725	26		76.384012	76.405458
12		539.325622	539.370747	27		150.250375	150.272626
13		538.651215	538.696951	28		75.711843	75.732089

(continued)

Table 6.7 (continued)

Serial number	Molecular structure	H/(Hartree)	G/(Hartree)	Serial number	Molecular structure	H/(Hartree)	G/(Hartree)
14		577.980846	578.029439	29		150.780201	150.801416
15		653.812552	653.864283				

6.4.1 Evolution of Methyl Radicals

It is learnt from the above analysis that the aliphatic hydrocarbon side chains in anthracite are dominated by short-chain methyl groups. Therefore, the chain reaction process of methyl radicals and hydroxyl radicals in anthracite was determined, as shown in (Eqs. 6.9–6.12), and the energy change of the radical reaction during the entire radical chain reaction was further calculated in combination with Table 6.7 (Zhang et al. 2024a, b), as shown in Fig. 6.18. The results showed that –OH will be attracted by C on the methyl radical to complement its position to become an alcohol hydroxyl group, and this step is a typical spontaneous exothermic reaction, whose reaction enthalpy becomes −314.70 kJ/mol. This implies that the bonding reaction between the methyl radical and the hydroxyl group can take place at the early stage of the spontaneous combustion of coal and become one of the reasons for the early exothermicity of the spontaneous combustion reaction of coal. After that, a large amount of –OH has been generated, and the latter is more oxidizing than oxygen, so the secondary alcohol hydroxyl group generated in the previous step will be attacked by –OH to become alcohol hydroxyl radical and water, which is a spontaneous exothermic reaction, and the enthalpy of the reaction became −49.41 kJ/mol. Furthermore, the alcohol hydroxyl radical generated in the reaction will continue to combine with –OH to form a stable carbonyl group and water. The reaction is spontaneous exothermic with an exotherm of −402.39 kJ/mol. The water molecules are expelled from the air as evaporates, and the secondary carbonyl radicals will continue to react under the action of –OH to form carbonyl radicals and water molecules, with an exotherm of 83.08 kJ/mol. Apparently, this reaction course confirmed the trend of the main microgroups of coal in the previous section. The aliphatic hydrocarbon side chains in coal were gradually oxidized to alcohol hydroxyl groups and further oxidized to carbonyl groups under the effect of oxygen. At the same time, the previous results showed that the decrease temperature of aliphatic hydrocarbons in anthracite is ca. 200 °C, while the carbonyl group undergoes a substantial increase only at roughly 250 °C, which means that the ability of methyl radicals in coal to generate water through a series of chain reactions would to increase after ~250 °C, and at the same time, the growth of the coal auto-ignition and exothermic effect caused by the reaction would to be reflected after about 250 °C. Obviously, the water content and exothermic amount of aliphatic hydrocarbon extracted coal at this stage should show a more obvious decreasing trend. In addition, the radical reaction of alcohol hydroxyl radical with –OH to generate water in the methyl chain reaction process, which is an easier reaction process, can also occur in the early stage of coal auto-ignition, which suggests that the moisture generation of ethylenediamine-extracted coal samples in the early stage of coal auto-ignition may also be weaker than that of other coal samples.

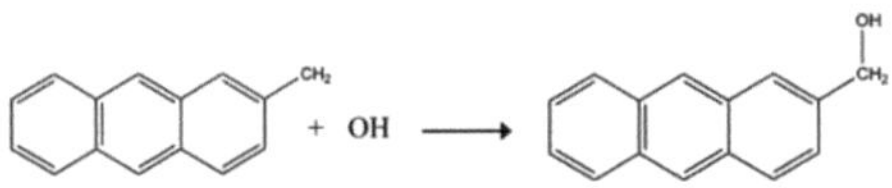

(6.9)

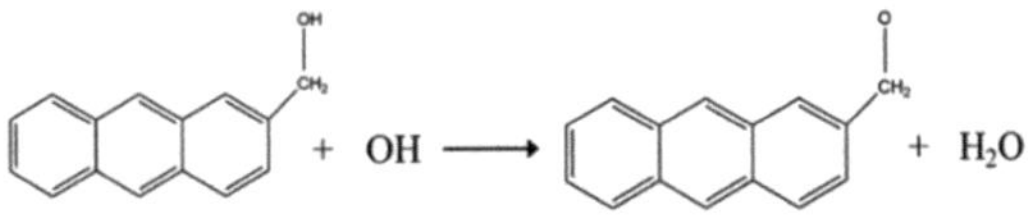

(6.10)

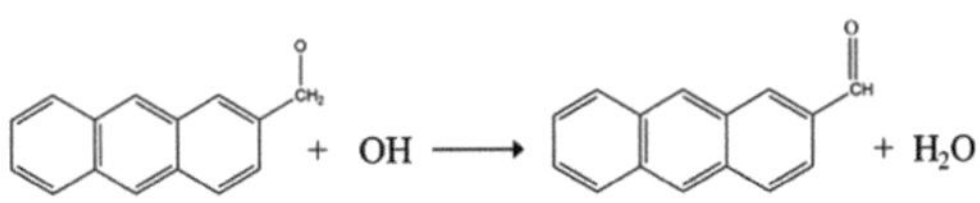

(6.11)

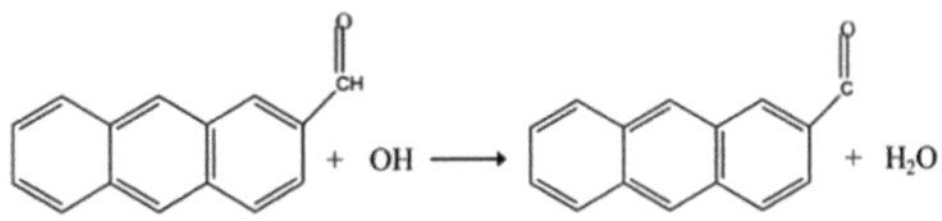

(6.12)

6.4.2 *Evolution of Methylidene Radicals*

As analyzed above, the development of aliphatic hydrocarbons in the side chain of bituminous coal is stronger than that of anthracite, and the side chain of the latter is mostly dominated by single-carbon methyl groups, while the former contains more two-carbon and even three-carbon groups, so the evolution of the sub-methyl radicals in the spontaneous combustion process of coal is based on the bituminous coal as the main structure. Based on this, the chain reaction history of the methylene radicals and hydroxyl groups in bituminous coal was determined, as shown in (Eqs. 6.13–6.16), and the energy changes of the radical reactions during the holistic free radical chain reaction process were further calculated in combination with Table 6.7, as shown in Fig. 6.19. The results showed that a hydrogen atom on the methylene group is taken away by the oxygen attack to form a methylene radical, and after encountering –OH, both of them undergo a combination reaction to form a more stable alcohol hydroxyl group, and this step is a typical spontaneous exothermic reaction with an exothermic amount of 319.06 kJ/mol. This is in line with the conclusion in Chap. 4 that the aliphatic hydrocarbons in coal are easily oxidized to hydroxyl groups by oxygen. Meanwhile, comparing the reaction heat of methyl group to generate alcohol

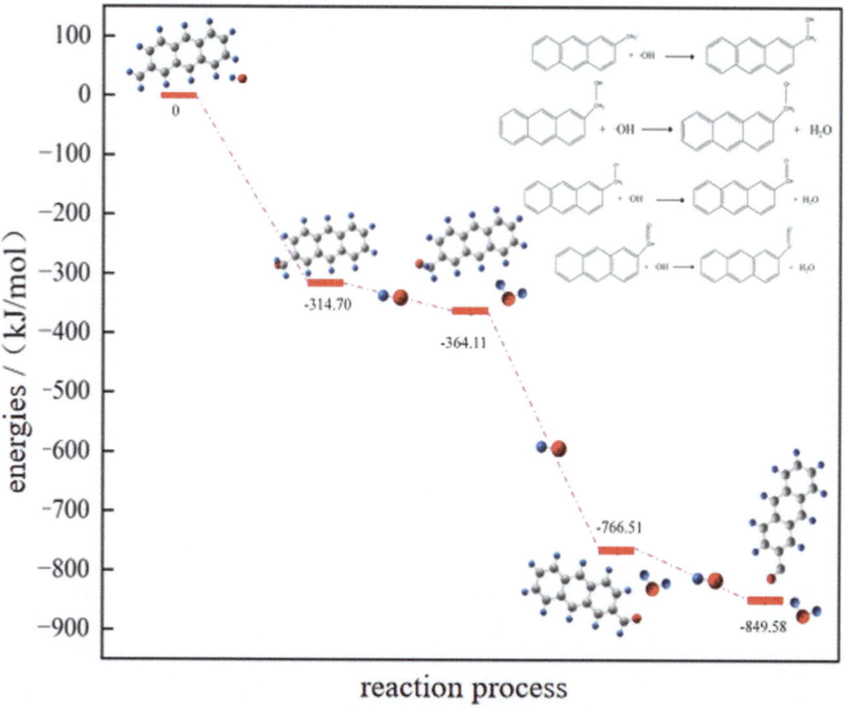

Fig. 6.18 Methyl radical reaction history and thermal effect in coal

hydroxyl group, it can be found that the reaction heat of the two is similar, so the conditions of occurrence are basically the same. After that, the generated alcohol hydroxyl group will be further oxidized by –OH to alcohol hydroxyl radical and release water, which is a spontaneous exothermic reaction with an enthalpy change of −60.37 kJ/mol. However, unlike methyl, the alcohol hydroxyl radical formed by methylidene is connected to $–CH_3$ molecule on the outer side, and therefore, the chemical bond between them is easy to break and form a methyl radical. The Gibbs free energy of this reaction was less than 0 and the enthalpy change was more than 0, which is a typical spontaneous heat absorption reaction, its heat absorption was 15.71 kJ/mol, while the methyl radical generated by the fracture was easy to capture the hydrogen atoms of the reaction fracture during the spontaneous combustion process to form CH_4, which is one of the main sources of methane at the late stage of the spontaneous combustion reaction of coal. Finally, meanwhile a large amount of –OH has been generated, the secondary carbonyl group generated will be oxidized by –OH to form carbonyl radical and water, with an exothermic amount of 83.18 kJ/mol. Observation of the entire reaction course revealed that the methylene radical was oxidized gradually by oxygen to carbonyl, and at the same time methane gas was released, and the secondary carbonyl group generated will be oxidized by –OH rather than by oxygen, and the holistic reaction The holistic reaction is a more obvious exothermic process. The whole reaction is a more obvious exothermic process, while the process of generating methane by molecular bond breaking is a non-spontaneous heat-absorbing reaction, which is more difficult to occur at the early stage of the coal auto-ignition reaction. Therefore, after the spontaneous combustion of coal was reached a certain stage, enough heat accumulation would cause the methylene in the coal to oxidize to form CH_4, which means that the CH_4 of bituminous coal may show a ineligible increase in the later stage of the reaction. In addition, because the aliphatic hydrocarbon content of the coal samples decreased after petroleum ether extraction, the generation temperature and amount of CH_4 in the petroleum ether-extracted coal samples may be lower than that in other coal samples.

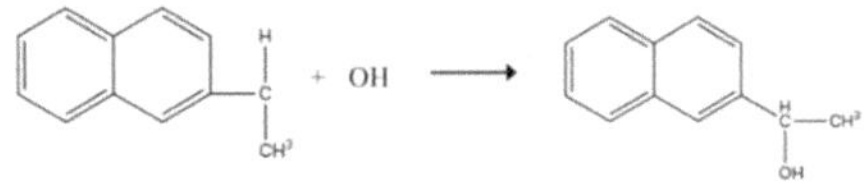

(6.13)

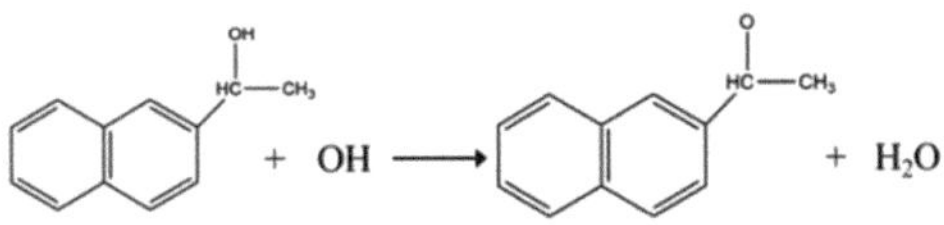

(6.14)

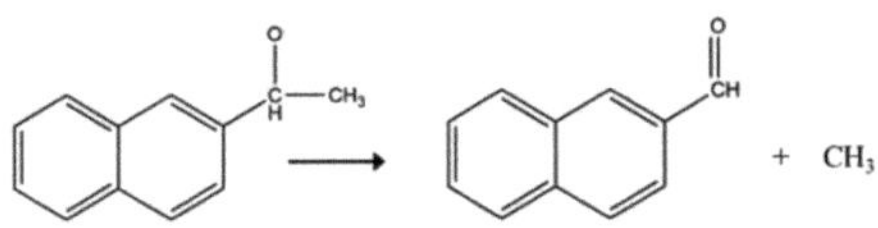

(6.15)

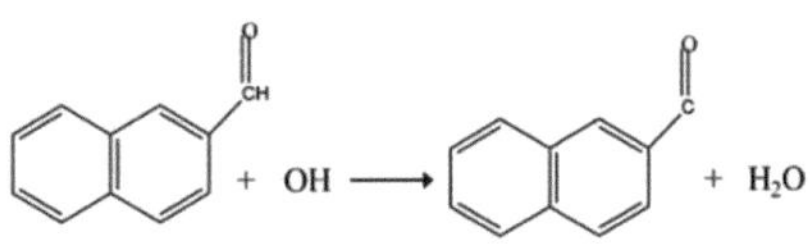

(6.16)

6.4.3 *Evolution of Carbonyl Radicals*

The above analysis showed that the chain reaction of either methyl, methylene or alcohol hydroxyl groups would eventually and gradually be converted to carbonyl groups, thus carbonyl groups were the most critical intermediate groups in the course of the spontaneous coal combustion reaction (Tian 2023). Carbonyl groups form carbonyl radicals when oxidized by oxygen (Zhang et al. 2024a, b). Based on the previous analyses and the results of the above studies, the chain reaction process of carbonyl radicals with hydroxyl radicals in bituminous and anthracite coal is determined as shown in (Eqs. 6.17–6.22), and the energy changes of the radical reactions during the whole radical chain reaction process were further calculated in combination with Table 6.7 as shown in Fig. 6.20. It can be seen that, similar to the reaction of alcohol hydroxyl radicals, the isolated electron on the carbonyl radical likewise attracted the isolated electron on the hydroxyl radical and ultimately trapped it to generate the carboxyl group. The entire reaction was a spontaneous exothermic

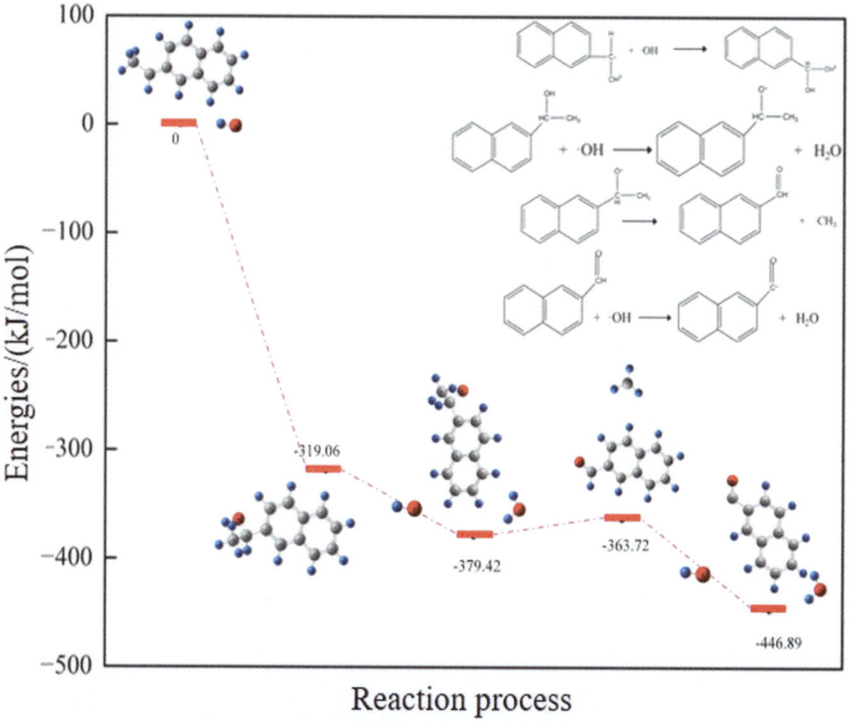

Fig. 6.19 Reaction history and thermal effect of methylene radicals in coal

reaction with exothermic quantities of 447.00 and 446.36 kJ/mol, respectively. In Chap. 4, Sect. 6.3 on the evolution of IR spectra of the main groups in coal, the conversion temperatures of the carbonyl group to the carboxyl group in coal took place around 250 and 300 °C, respectively, which revealed that the contribution of the exothermic effect of this reaction was mainly concentrated in this stage, i.e., the pre-start of the decomposition reaction. Obviously, the energy released from the evolution of carbonyl to carboxyl group provided a certain initial impetus for the decomposition of coal macromolecules and the precipitation of volatile matter.

After the generation of the carboxyl group, OH will continue to react with the carbonyl group instead of oxygen, and the H atom on its alcohol hydroxyl group was be taken away to form a water molecule, which itself evolves into a carboxyl radical. The reaction was spontaneous and exothermic with enthalpy changes of 41.08 and 41.23 kJ/mol, respectively. Next, the carbon atoms on the carboxyl radical attached to the aromatic ring were broken to generate aromatic radicals and CO_2, which is a non-spontaneous and absorptive reaction with heat absorption of only 20.95 and 21.50 kJ/mol, respectively. In general, the oxidation of carbonyl radical to generate CO_2 is an exothermic reaction, and the exothermic amount of the reaction was 467.13 and 466.74 kJ/mol, respectively. For anthracite, owning to the negligible content of carboxyl group, the CO_2 content in the reaction process basically came from the

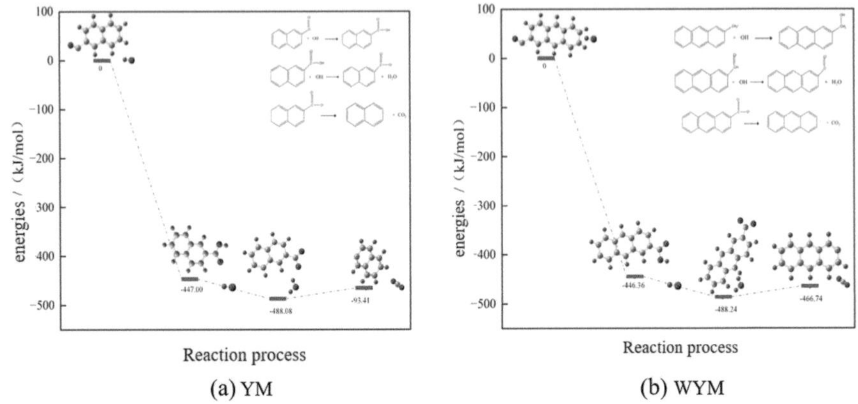

Fig. 6.20 Reaction history and thermal effect of carbonyl radicals in coal

oxidation of carbonyl radical. Therefore, the generation of CO_2 gas from anthracite after NMP extraction may distinctly lag behind that of other coal samples.

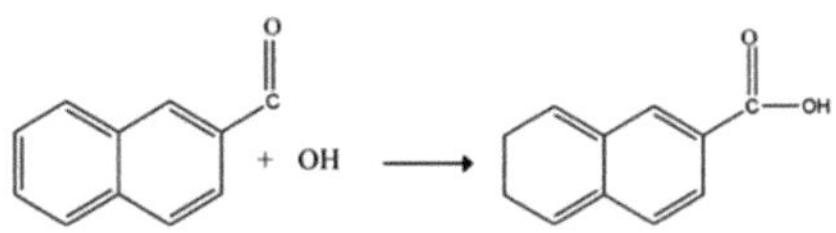

(6.17)

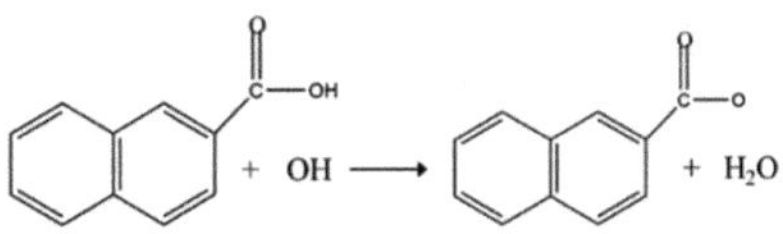

(6.18)

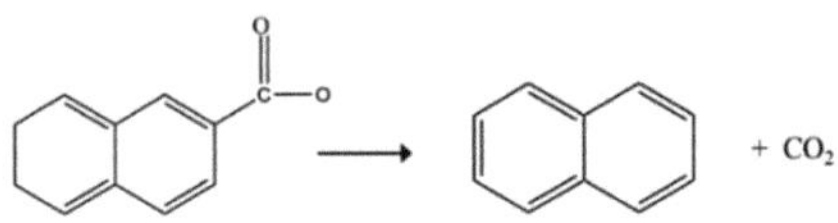

(6.19)

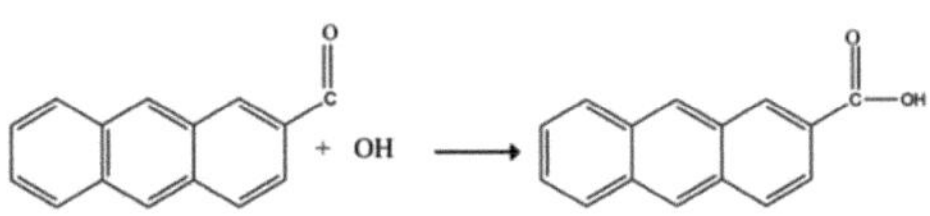

(6.20)

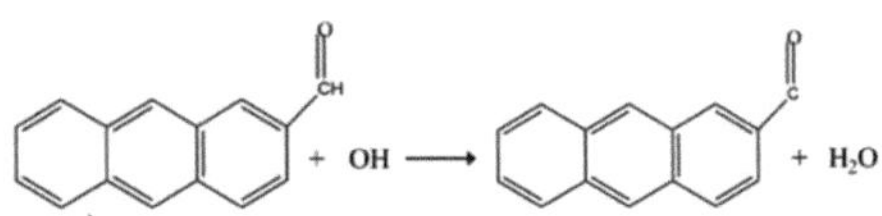

(6.21)

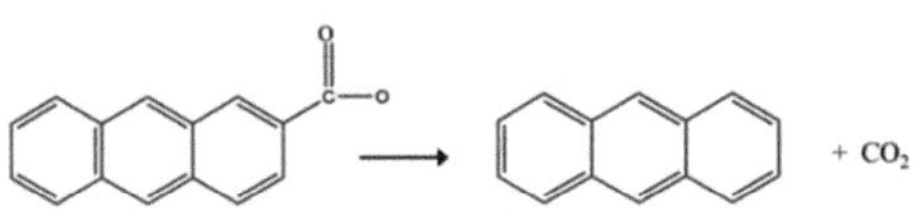

(6.22)

6.4.4 Coal Spontaneous Combustion Gas Generation Mechanism and Hydroxyl Chain Self-promoting Oxidative Coal Spontaneous Combustion Mechanism

From the above analysis, it could be found that for the reactive groups in YM and WYM, whether they are methyl radicals, methylene radicals or alcohol hydroxyl radicals, their chain reaction course will eventually oxidize to carbonyl groups or carbonyl radicals, which were further oxidized to yield carboxyl radicals. Therefore, the carbonyl radical was a key link that is difficult to be bypassed in the free radical chain reaction mechanism of spontaneous coal combustion. Accordingly, the key radical chain reaction and the main gas generation mechanism of coal autogenous combustion with carbonyl group or carbonyl radical as the key group were established, as shown in Fig. 6.21.

Figure 6.21 shows that during the spontaneous combustion of coal, O_2 firstly attacked the active sites on the structure of coal macromolecules, i.e., the active groups such as alcohol hydroxyl, methyl, methylene, and carbonyl groups, and seize the H atoms on the reactive sites to form OH, which was easy to evolve into OH

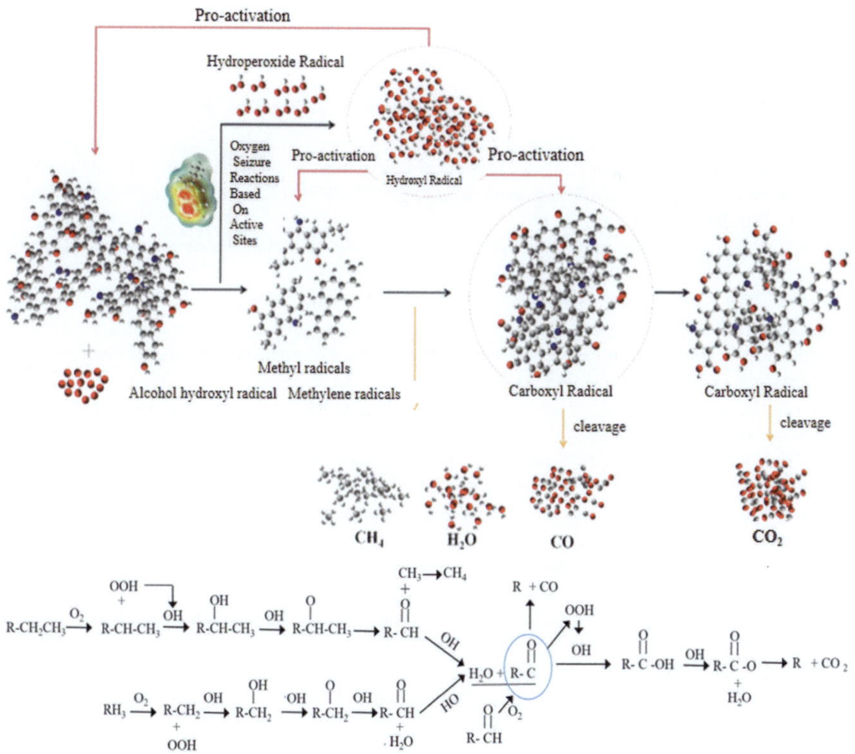

Fig. 6.21 Hydroxyl chain self-promoting oxidized coal auto-ignition mechanism

with high oxidative activity to continue to participate in the reaction. Afterwards, the reactive group evolved into a reactive radical after capturing an H atom, for example, methyl and methylene radicals will further reacted with OH and O_2 in a chain reaction to release H_2O and CH_4, and finally evolve into a carbonyl group. Then, the carbonyl group was repeatedly oxidized by oxygen to become carbonyl radical and became the main source of CO during spontaneous coal combustion. The carbonyl radical continued to complex with –OH to become a carboxyl group and further evolved into a carboxyl radical, ultimately generating an amount of CO_2. OH was therefore the key reactive small molecule group in the chain reaction process of coal auto-ignition, which was generated by oxygen seizing the H atoms of the reactive group and continues to take part in the chain reaction of the remaining free radicals, and the carbonyl group is the key group that carried on the chain reaction of the process of coal auto-ignition, which is divided into the carbonyl or the carbonyl radical as the dividing line, the generation of gas products can be divided into three stages. Before the oxidation of carbonyl group, the main product of the oxidation of the main group in coal was H_2O, and the by-product was CH_4. After the oxidation of carbonyl group or carbonyl radical, the product were CO, and when the carbonyl group was further oxidized to carboxyl group or carboxyl radical, the product was CO_2.

References

Bi HD, Wang Y, Zhao X et al (2024) Theoretical calculation of alkaline hydrolysis of heterobifunctional reactive dye-cellulose compounds. J Silk. 61(04):9–17. https://doi.org/10.3969/j.issn.1001-7003.2024.04.002

Deng J, Li YQ, Zhang YT et al (2020) Effects of hydroxyl on oxidation characteristics of side chain active groups in coal. J China Coal Soc 45(1):232 240

Fukui K (1985) Frontier orbitals and chemical reactions. Science Press, New York. https://doi.org/10.2307/1689733

Ge LM, Li JW (2003) Evolution of functional groups in low-temperature oxidized Shenfu coal. J xi'an Univ Sci Technol 23(2):187–190

Guo ZG, Zhang J, Zheng BH et al (2023) Evolution characteristics and microscopic dynamics mechanism of recrudescence behaviors of preoxidized coal. J Saf Environ 23(08):2644–2652

Li HL, Yang M, Sun J et al (2021) Synthesis, crystal structure and theory studies of a cobalt complex with anion-water cluster. J Synth Cryst 50(12):2293–2299

Li YQ, Ou LF, Li X et al (2024) Study on thermogravimetric characteristics and key functional groups of coal spontaneous combustion. J Saf Sci Technol 20(02):118–124

Li YM, Advanced Organic Chemistry (2018) Reactions, mechanisms and structures. Chemical Industry Press

Li X (2022) Study on the reaction path and inhibition mechanism of coal oxidation of hydroxyl radicals at low temperature. Tianjin University of Technology, Tianjing. https://doi.org/10.27360/d.cnki.gtlgy.2022.000042

Lou HZ, Jia TG (2019) Study on thermal characteristics of coal spontaneous combustion in gas atmosphere by TG-DSC coupling techniques. China Saf Sci J 29(11):77–82

Lu WD, Wang JR, Deng CB et al (2017) Experiment research on inhibitors of coal spontaneous ignition based on activation energy index. Mod Min 10:45–89

Luan T, Kuang XH, Wang W et al (2024) Research on progress of quantum computing simulation of physical systems. J Front Comput Sci Technol 12:1–11

Qian YY, Zhang SP, Gao JJ et al (2015) Fabrication of graphene-based nanomaterials by non-covalent functionalization and their applications. Chemistry 78(06):497–504

Sun W, Jiang CX, Liang JJ et al (2023) The DFT study on the reaction of Edaravone with hydrogen peroxide radical in aqueous solution. J Nankai Univ 56(06):94–104

Tian G (2023) Study of the kinetics characteristics and the products formation laws of spontaneous combustion of coal-polyurethane binary system. China University of Mining and Technology, Beijing. https://doi.org/10.27624/d.cnki.gzkbu.2021.000070

Wang DM, Xin HH, Qi XY et al (2014) Mechanism and relationships of elementary reactionsin spontaneous combustion of coal: the coal oxidation kinetics theory and application. J China Coal Soc 39(8):1667–1674

Wang X, Zhang H, Wang K et al (2024) Mechanistic study of the effect of differential electron density characteristic groups on the nitrogen migration transformation in coal. Clean Coal Technol 43:1–10

Xia P (2023) Density functional theory study on composites catalyzing CO_2 reduction and CH_4 partial oxidation. Wuhan Institute of Technology, Hubei. https://doi.org/10.27727/d.cnki.gwhxc.2022.000395

Xu JC (2001) Determination theory of coal spontaneous combustion zone. China Coal Industry Publishing House

Zhang YN (2012) Study on the microcosmic characteristics and macro parameters in the process of coal oxidation and spontaneous combustion. Xi'an University of Science and Technology, Shaanxi

Zhang H, Tang L (2020) Study on reaction mechanism of p-type dopant Cp2Mg in MOCVD gas phase. CIESC J 71(07):3000–3008

Zhang W, Zang LY, Li J (2024a) Application and progress of quantum chemical methods in the mechanism of spontaneous coal combustion. Sci Technol Rev 42(04):73–83

Zhang YB, Zhang YT, Deng J et al (2024b) Evolution characteristics of key groups of coal oxidation exotherm and their thermal effects under oxygen-depleted condition. J China Coal Soc 25:1–13

Zhang R (2024) Preparation and performance of catalysts for partial oxidation of methane under mild conditions. Beijing University of Chemical Technology, Beijing. https://doi.org/10.26939/d.cnki.gbhgu.2023.000532

Zhao GL, Li WT et al (2024) A theoretical investigation of the adsorption of three different microplastics and cefradine. Mar Environ Sci 43(02):227–234

Zhao Y, Li X, Huang JK et al (2024) Mechanism of heterogeneous reduction of NO over graphite-supported single-atom Fe catalyst: DFT study: DFT study. J Fuel Chem Technol 52(05):717–724

Zhao MF (2023) Experimental study on the inhibitory characteristics of low metamorphic coal in the critical temperature region. Liaoning Technical University, Liaoning. https://doi.org/10.27210/d.cnki.glnju.2023.000287

MIX
Papier aus verantwortungsvollen Quellen
Paper from responsible sources
FSC® C105338

If you have any concerns about our products,
you can contact us on
ProductSafety@springernature.com

In case Publisher is established outside the EU,
the EU authorized representative is:
Springer Nature Customer Service Center GmbH
Europaplatz 3, 69115 Heidelberg, Germany

Printed by Libri Plureos GmbH
in Hamburg, Germany